W9-CAB-764

ADVANCE PRAISE FOR *The Prism and the Rainbow*

---

"Dr. Martin has written an accessible and interesting introduction to evolution for teenagers who are afraid that the basic theories of modern biological science are incompatible with their Christian faith. Martin shows why this is not so and why their commitment to the belief that God is creator should stimulate rather than discourage their study of natural science. The book meets a real need in Christian education."

—Dr. David C. Steinmetz, Duke University Divinity School, Amos Ragan Kearns Professor of the History of Christianity, general editor of *Oxford Studies in Historical Theology,* former president of the American Society of Church History

"Professor Martin has produced a concise yet comprehensive exposition of the fundamental issues in the current debates between (and among) Christians and evolutionary biologists on creationism, intelligent design, and theistic evolution. It addresses all the key issues, both biological and theological, in a form and language easily accessible to Christian laity, including teens, and evaluates them fairly and with erudition. I would recommend this book without hesitation for use in both parish and university settings."

—Dr. Jarvis Streeter, California Lutheran University, Professor of Religion, author of *Human Nature, Human Evil and Religion: Ernest Becker and Christian Theology*

"Professor Martin has a very good understanding of the relationship of science to religion. He applies that understanding effectively in this sensitive and sensible discussion of evolution and Christian faith."

—Dr. John F. Haught, Georgetown University, Senior Fellow, Science and Religion, Woodstock Theological Center, author of *God after Darwin, Science and Religion,* and *What Is God?*

"*The Prism and the Rainbow* is a wonderful introduction to the evolution-creation debate by someone who is a practicing scientist but who is also deeply understanding of and sympathetic to those for whom their Christian faith is all important. It is written for young people and I believe would find eager response from them. I think it would be a wonderful book to use in discussions both at school and at church. I could see it having a wide readership and doing much good in the direction of reconciling those who want to run with the hare of science and hunt with the hounds of religion."

—Dr. Michael Ruse, Florida State University, Lucyle T. Werkmeister Professor of Philosophy and Director of the Program in History and Philosophy, author of *Can a Darwinian Be a Christian? Darwin and Design,* and *The Evolution-Creation Struggle*

"Teenagers need a place where they can get straight talk and reputable answers about evolution, all the more so if they come from religious backgrounds that prepare them to mistrust what they might be told. Jody Martin provides a safe place for them. He is a first-class, accomplished scientist and a dedicated Christian, and his is a perspective well worth listening to."

—Dr. Kevin Padian, Department of Integrative Biology, University of California, Berkeley, co-editor of *Encyclopedia of Dinosaurs*

"A concise and clear explanation of how science and faith are not at war with each other but instead can complement our understanding of the world that God created."

—Rob Seitz, CEO, Applied Genomics, Inc., Huntsville, Alabama

"Dr. Martin explains why Christians should have confidence in their faith while accepting the science of evolutionary biology. There is no conflict between Christianity and evolution, and Dr. Martin explains why with extraordinary clarity. This book should be required reading for all Christian teens."

—Dr. Linda Silver, President and CEO, Great Lakes Science Center

"In this balanced and accessible discussion of evolutionary theory and Christianity, Joel Martin offers his readers a thoughtful way beyond the usual rhetoric of "conflict" between Darwinism and religious belief. Martin's critique of creationism and Intelligent Design is sharply focused without being strident, while his own constructive synthesis is imaginative and insightful. This brief book provides a splendid example of how a respectful dialogue between science and spirituality can enrich both perspectives."

—Dr. Andrew Lustig, Davidson College, Davidson, North Carolina, Holmes Rolston III Professor of Religion and Science, founding co-editor of *Christian Bioethics*

"Joel Martin's marvelous book echoes the sentiments of Charles Darwin himself, who reminded readers of *The Origin* that one cannot be too learned 'in the book of God's word, or in the book of God's work.' It is a powerful testament to the compatibility of faith and reason, and demonstrates with remarkable clarity that the conflict between religion and evolution is a false and contrived choice that does justice to neither science nor religion."

—Dr. Ken Miller, Brown University, author of *Finding Darwin's God, Only a Theory: Evolution and the Battle for America's Soul* and (with Joe Levine) three popular high school biology textbooks

"*The Prism and the Rainbow* provides a very effective introduction to arguments—some good, some bad—swirling around evolution and its critics in contemporary churches and their social context. A careful, critical reading of it could help young people get off to a good start in dealing with issues that trouble many people today."

—Dr. Dallas Willard, University of Southern California, Professor of Philosophy, author of *Hearing God, Knowing Christ Today, The Divine Conspiracy,* and *Renovation of the Heart*

# *the* Prism *and the* Rainbow

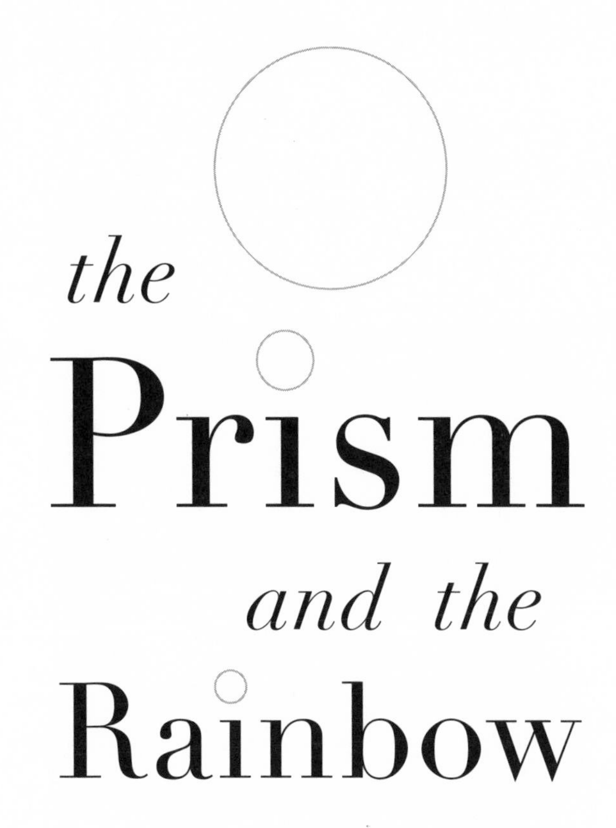

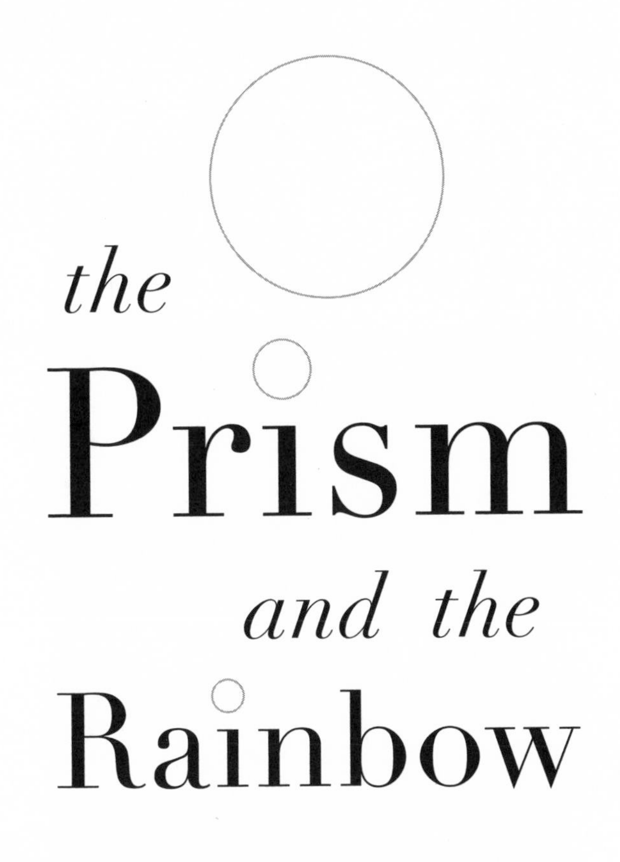

# *the* Prism *and the* Rainbow

*A Christian Explains Why Evolution Is Not a Threat*

Joel W. Martin

THE JOHNS HOPKINS UNIVERSITY PRESS
BALTIMORE

Printed in the United States of America on acid-free paper
2 4 6 8 9 7 5 3 1

The Johns Hopkins University Press
2715 North Charles Street
Baltimore, Maryland 21218-4363
www.press.jhu.edu

Library of Congress Cataloging-in-Publication Data

Martin, Joel W., 1955–
The prism and the rainbow : a Christian explains why evolution is not a threat /
Joel W. Martin.
p. cm.
Includes bibliographical references and index.
ISBN-13: 978-0-8018-9478-7 (hardcover : alk. paper)
ISBN-10: 0-8018-9478-6 (hardcover : alk. paper)
1. Evolution (Biology)—Religious aspects—Christianity. 2. Creationism.
3. Religion and science. I. Title.
BL263.M418 2010
231.7'652—DC22 2009030921

A catalog record for this book is available from the British Library.

Book design by Kimberly Glyder Design

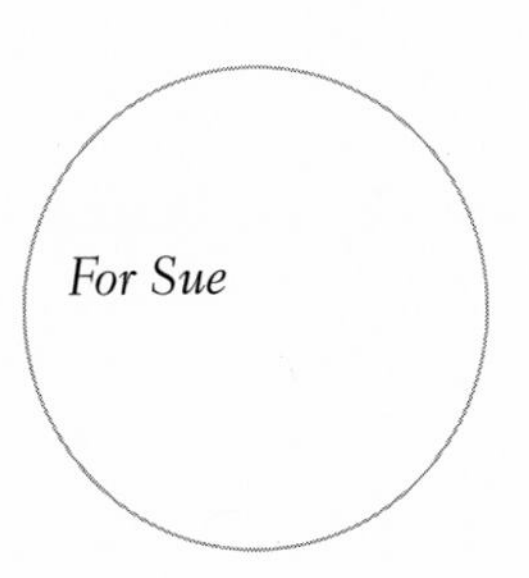
*For Sue*

*An intelligent mind acquires knowledge,*
*and the ear of the wise seeks knowledge.*

PROVERBS 18:15

*Let knowledge grow from more to more,*
*But more of reverence in us dwell;*
*That mind and soul, according well,*
*May make one music as before.*

ALFRED, LORD TENNYSON, "IN MEMORIAM A.H.H."

# Contents

Introduction...PAGE 1

**1** By the Numbers...PAGE 9

**2** The Prism and the Rainbow...PAGE 15

**3** The Flat Earth Society...PAGE 21

**4** Of Serpents and Certainty...PAGE 27

**5** The Nature of Science...PAGE 31

**6** What Does "Theory" Mean?...PAGE 39

**7** What Is Evolution?...PAGE 47

**8** What Is Creationism?...PAGE 55

**9** What Is Intelligent Design?...PAGE 63

**10** Is There Evidence Supporting Intelligent Design?...PAGE 69

**11** Human Arrogance...PAGE 81

**12** In the Beginning...PAGE 87

**13** The Unnecessary Choice...PAGE 93

**14** What Are We to Believe?...PAGE 99

Epilogue...PAGE 103

**ACKNOWLEDGMENTS**...PAGE 107

**APPENDIX**: *Major Christian Denominations and Their Stance on Science, Evolution, and Creationism/Intelligent Design*...PAGE 111

**NOTES**...PAGE 131

**GLOSSARY**...PAGE 151

**RECOMMENDED FURTHER READING**...PAGE 157

**HELPFUL WEB SITES**...PAGE 161

**INDEX**...PAGE 163

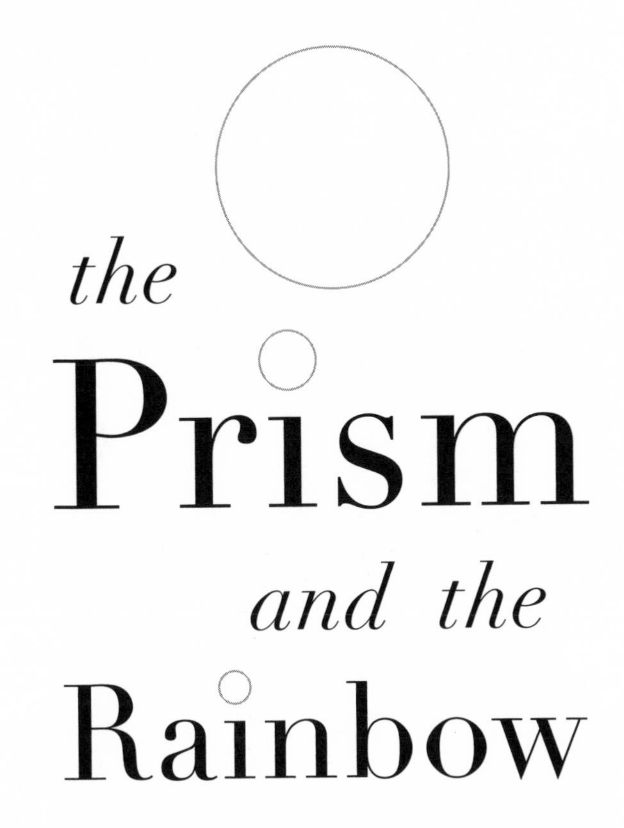

# *the* Prism *and the* Rainbow

# Introduction

*He said to him, "'You shall love the Lord your God with all your heart, and with all your soul, and with all your mind.' This is the greatest and first commandment."*

MATTHEW 22:37–38

*God is . . . the embodiment of the laws of the universe.*

STEPHEN HAWKING

Several years ago, some students who attended the same church that I attend stood up in their high school biology class to announce that they were against the teaching of evolution because they thought that it went against their religious beliefs. It probably took a considerable amount of courage to get up in class and mount a protest. Clearly, they felt strongly about their position. The students, all of whom were intelligent, motivated, and caring individuals, were moved to speak out against something they perceived as non-Christian.

They were surprised, more than a little confused, and probably hurt to realize later that leaders in their church, our church, did not support their stance. Indeed, not only was this not the stance of that particular congregation, it also was not the position of the church (the Presbyterian Church USA)[1] nationwide. Nor is it the position of the majority of Christians worldwide or even in the United States, where anti-evolution sentiment is so high (see chapter 1 and the appendix). The reason is fairly simple: according to most Christian theological interpretations, evolution, and the abundant evidence for it, are also part of God's world—just as is the case with geology, physics, astronomy, chemistry, and all of the other fields of science. The students' confusion caused several of the church leaders to begin to question how, when, and what we teach our children, and how we equip them with the knowledge and faith to navigate successfully as intelligent, practicing Christians in today's world.

This book is written for anyone who might be similarly confused about the relationship between science and faith. I have kept it purposefully brief so that it can be read quickly. As a parent and as a youth worker, I know how busy students are today. But I am not writing just for students. I am also writing for the parents of those students and for other adults, partly as a response to the many requests I have received over the last two decades.

There is a vast amount of literature available on the subject of "creationism vs. evolution," some of which is excellent and some of which is not.[2] In light of all that literature, and particularly because of the perceived divide between science and faith, it's important that readers know at least a little about my background in both science and religion.

I am a Christian, and I grew up in a Presbyterian home in North Carolina and (later) Kentucky. I am also an ordained elder in the Presbyterian Church USA. For a little more than seven years I was a youth advisor at a Presbyterian Church in southern California, working with middle and sometimes high school students. After a short break, I worked for the next eight years more exclusively with the high school students, and I continue to teach Sunday school and confirmation classes to high school students today.

In terms of my background and knowledge of the Bible, I do not consider myself a "Bible scholar," mostly because I do not read Aramaic, Hebrew, Greek, or any other of the original languages and early translations of the Bible. I also have no formal training (no advanced college degrees or seminary classes, for example) in the history, translation, or interpretation of the Bible. Thus, like many of you, for a more in-depth understanding of the Bible I have to rely pretty heavily on various English translations of the Bible, on English-language books about its history and meaning, and on my pastor and other church leaders whose backgrounds and training are much stronger in these areas. I have done a lot of reading of the various translations, interpretations, analyses, and summaries that are available concerning the Bible and the creationism/evolution controversy, but possibly no more than many other people who are truly curious about the Christian faith and its relationship with science over the years.

So while there are many people in our church and elsewhere who are more actively involved in a variety of ministries, more widely read in the field of Holy Scriptures, and more educated with regard to the Bible itself, I am at least aware of the major issues. I make this point primarily to counter the simplistic argument, put forth by several (though not all) creationists and proponents of "intelligent design," that those who understand and teach evolution and other sciences are all atheists. That view is simply false.

On the scientific side, my credentials are stronger.[3] I have an advanced degree in biology, I work as a curator at a major natural history museum, and I am an adjunct professor at two universities (University of California, Los Angeles, and the University of Southern California), where I have taught several classes. Most relevant to this book is a class that I taught with two other professors in the early 1990s titled "Population Genetics and Evolution," a course for upper division undergraduate students and graduate students at USC. Like nearly all museum scientists, I publish scientific articles, including studies on evolutionary topics, and like all students of natural history I am deeply concerned with and involved with understanding our natural world. My research specialty is a field of science called systematic biology, which is the study of the relationships among organisms, a discipline that is a part of the larger field of evolutionary biology.

This might seem like an odd combination—a Christian who is also a scientist heavily involved with the study and teaching of evolution. But it really isn't. Polls have shown that a large number of scientists have a strong faith in one religion or another and that the majority of religious leaders have no quarrel with science.[4] The idea that all scientists are atheists or that all Christians are fundamentalists is simply a myth, one that has been intentionally perpetrated in some instances. Some scientists are atheists or agnostics—so are some plumbers, some professional athletes, some nurses, some politicians, and so on—but not all of them. Some Christians are fundamentalists, but not all of them. Sweeping generalizations like that are usually wrong, and they most certainly are wrong in this case.

Given that so much has been written about the evolution and creationism controversy, and the larger issue of religion and science and how (or if) they should interact, I hesitate to write yet another book on the subject, no matter how short. To some extent, the issue can be resolved by two very simple statements:

1. Religion is not science and should never masquerade as such.
2. Science is restricted to observing and testing phenomena in the natural world around us and should never be used to argue for or against a particular faith or set of religious beliefs.[5]

By definition, then, these two areas of human endeavor, science and religion, address different issues, each using distinctive methods of inquiry, and there should be no "debate." But that is not the end of the story, and confusion still seems to reign. Often discussions on the topic have been hostile; Baptist minister and Harvard professor Peter Gomes describes modernity as "a series of guerrilla wars between an egocentric Christianity and an arrogant secular science, neither of which is prepared to concede to the other, neither of which can achieve absolute and unambiguous victory, and neither of which is prepared to take any prisoners."[6]

Arrogance continues to play a large role on both sides of the issue, with some (not all) vocal devotees on one side claiming to know the mind of God and some (not all) on the other side claiming that God does not exist. The majority of people adopt a more humble view, and many see common ground that is fertile for discussion with potential benefits to both science and theology.[7] People in general are interested and curious about the topic, which is always a good thing, and wondering how to resolve what is so often seen as a dispute.

Thus, I am convinced that there is a need for a cogent explanation of how Christians should view science, including evolution, written by someone who is both a scientist with an understanding of evolution and a Christian. It is difficult to approach the topic without also getting into the question of fundamentalism and literalism in biblical interpretation, and so I will touch on those themes in order to make a couple of points that I think are important. Mostly, I want to convey to readers that the debate over science and faith, and specifically over evolution and creationism, is seen as a problem mostly to those who lack a clear understanding of the other side, and especially to those who perceive that there is something anti-religious or anti-Christian about science and the evidence that abounds in the world around us. Despite the fact that many people would like to think otherwise, this is not a simple case of right vs. wrong, good vs. bad, or theistic vs. atheistic worldviews.

The verses I used to introduce this section (Matthew 22:37–38) make it abundantly clear that heart, soul, and mind are all to be seriously involved as part of the Christian life. Yet it seems to me that use of the mind has been curiously absent from many Christian statements and discussions concerning science and evolution. My premise in this book is that God exists. I will argue that science and faith are clearly distinct yet not incompatible. That biological evolution has occurred and continues to occur. That evolution is a valid, exciting field of science that does not threaten the Christian worldview. That far too much fear and acrimony have been voiced. And that within the human heart there is room for, and a need for, both a firm understanding of science and the peace and hope that accompany faith in God.

# 1
# By the Numbers

*He determines the number of the stars;*
*he gives to all of them their names.*

PSALM 147:4

*Not everything that counts can be counted, and not everything that can be counted counts.*

ANONYMOUS

In my introduction, I made the claim that most Christians, worldwide and in the United States, are accepting of modern science, and specifically evolutionary biology, as being fully compatible with their faith.

Is this really true? And on what evidence would I base such a statement? Nearly everyone is aware that there has always been a large number of well-known and highly respected scientists who are also committed Christians (from Galileo and Isaac Newton to Theodosius Dobzhanksy, Ronald Fisher, Francisco Ayala, Kenneth Miller, Francis Collins, Simon Conway Morris, and many others). And there have

TABLE 1. Position on evolution of selected Christian organizations or denominations (arranged by decreasing estimated membership in the United States).

| DENOMINATION OR MOVEMENT | ACCEPTING OF EVOLUTION AS BEING COMPATIBLE WITH THEIR FAITH | | POSITION UNKNOWN OR UNCLEAR |
|---|---|---|---|
| Roman Catholic | yes | | |
| Southern Baptist Convention | | no | |
| United Methodist | yes | | |
| National Baptist Convention USA | | no | |
| Church of God in Christ | | no | |
| Evangelical Lutheran Church in America | yes | | |
| African Methodist Episcopal Church (AME) | yes* | | |
| Church of Christ | | | unclear |
| International Circle of Faith | | no | |
| Anabaptists | | | unclear |
| Presbyterian Church (PCUSA) | yes | | |
| Calvary Chapel | | no | |
| Church of God (Cleveland) | | no | |
| Assemblies of God | | no | |
| Lutheran Church-Missouri Synod | | no | |
| Episcopal Church | yes | | |
| Greek Orthodox Archdiocese of America | yes | | |
| United Church of Christ | yes | | |
| Seventh-Day Adventists | | no | |
| Presbyterian Church in America | | | unclear |
| The Vineyard | | | unclear |
| International Church of the Foursquare Gospel | | no*+ | |
| New Apostolic Church | | no*+ | |
| TOTAL MEMBERSHIP (IN MILLIONS) | 94.05 | 45.85 | 8.5 |

**NOTE:** Totals at bottom reflect estimated membership numbers in the United States. See appendix for details of how these estimates were derived. *Assumed position based on views of other Methodist (for AME), Baptist (for NBC USA), or Pentecostal groups (for New Apostolic, ICFG, and COGIC). + = no current estimate for membership number at this time.

always been many religious leaders who have fully embraced evolution as part of God's world (Billy Graham, Pope John Paul II, John Polkinghorne, Arthur Peacocke, John Haught, to name just a few). But isn't opposition to evolution a position held by most Christians? It turns out that it is not.

In table 1, I have summarized the information (for details, see the appendix) that I could find on the major Christian denominations in the United States and their stated position on science and faith, creationism and evolution. The denominations or movements are listed in decreasing order of the estimated number of members. When looking at the table, we must keep in mind a large number of caveats and considerations. Let me list just five of these here.

First, it is difficult to find accurate estimates of membership numbers for many religious groups and organizations. Some groups do not maintain membership figures, some use attendance figures, and some have no central governing body or headquarters. The various sources I used for membership estimates, mostly books and web sites (see the appendix), differ in their estimates and even in which groups are included; a "large" group mentioned in one volume might not be included in another or might appear under a different name.

Second, not all denominations are listed here or anywhere else (there are said to be more than 1,200 Christian denominations or organizations in the United States alone). Denominations have come and gone over the years, especially in the United States. Names have changed, groups have merged, and there is apparently no single source that tracks all of these changes, although several do a respectable job of maintaining lists of U.S. denominations, and I have consulted these.

Third, not all groups have a statement or position on what they believe as concerns science, faith, and evolution. Some offer clear statements about their beliefs on this topic while others do not, and some offer statements that are slightly ambiguous.

Fourth, many Christians in the United States attend large "megachurches," several of which are nondenominational (see the appendix). This unknown but undoubtedly significant number of Christians probably is not represented in the membership counts of mainstream denominations.

Fifth, and perhaps most importantly, we should never assume that all members of a given denomination agree with, or are even aware of, the "official" position of the group to which they belong. Such statements, if they exist, usually are not meant to be binding on, might not reflect the opinions of, and might not even be known by the general membership of the group. I know Presbyterians who are creationists and Baptists who are not. Any large group will contain a diversity of beliefs and opinions. Within my own denomination (PCUSA)—and in fact even within the relatively small number of members in my own local church—it is easy to find quite divergent views on a wide range of theological issues. But since we cannot survey all of the roughly 2.2 billion Christians worldwide, for now I have relied on statements by the governing bodies and spokespersons for these groups.

Table 1 and the appendix are at best generalizations. Yet even with all of the above qualifications noted, it seems clear that acceptance of evolution is a majority, and not a minority, view among Christians.

If these numbers are accurate, or even if they are just a rough approximation, then we should ask: Why would this be? Why is it that most Christians believe that modern science (including evolution) is compatible with their faith, that it even strengthens and enriches their faith? One answer to this question lies, I believe, in the following chapters.

# 2

# The Prism and the Rainbow

*Ask, and it will be given you; search, and you will find; knock, and the door will be opened for you.*

MATTHEW 7:7

*Nature is the living, visible garment of God.*

JOHANN WOLFGANG VON GOETHE

As a child I learned that rainbows were a sign from God. I still see them that way. There are few things in this world so surprisingly and shockingly beautiful as a full-blown rainbow, and each time I see one I am momentarily amazed and thrilled. To be reminded of God every time that I see one, a holdover from those early years in Sunday school, is also, I think, not a bad thing. Indeed, there are days when I wish there were more reminders like that out there.

The association of rainbows with God is easy to trace. Although rainbows are also mentioned in the books of Ezekiel (1:28) and Revelation (4:3, 10:1), they are best known from Genesis, chapter 9, verses

11–17, where God is reassuring Noah that he will never create another deluge and that the rainbow is the sign of this promise:

> God said, "This is the sign of the covenant that I make between me and you and every living creature that is with you, for all future generations. I have set my [rain]bow in the clouds, and it shall be a sign of the covenant between me and the earth." (Genesis 9:12–13) [1]

It's a wonderful story, one that is extremely reassuring to a child afraid of lightning and thunder and wondering when the rain is going to stop. Additionally, it's easy to see how something as colorful, mysterious, and inexplicable as a rainbow simply had to be the work of God in the eyes of people who were not acquainted with the physics of light and water. Even for people today who fully understand the physics behind rainbows it is easy to think of the hand of God behind such a wondrous spectacle. In fact, it is hard for many people (and I am one of them) to conceive of a world that does not have God in it when confronted with such unbridled glory.

Eventually, I learned how it all works, how the different colors are generated when sunlight interacts with water in the atmosphere, which is also quite wonderful. We have Sir Isaac Newton to thank for this, as he worked it out back in the late seventeenth century. At the time of Newton's experiments, people believed that color was composed of mixtures of light and darkness and that a prism somehow added color to the light. Newton made a small hole in his window shutters to allow a narrow ray of sunlight in, and then he placed a prism close to the window. The ray of sunlight hit the prism and was divided (refracted) into a rainbow of colors on the far wall. He then used a second prism, placed upside down just behind the first prism, to show that the colors could be "reassembled" back into white light, proving clearly that light alone was responsible for the color. The prism did not create the color, it simply revealed it.

Newton's work also explained the regularity to the colors in the rainbow: they are always arranged in the same order—violet, blue, green, yellow, orange, red. When sunlight passes through a prism, it is refracted (bent) and the colors are separated because the different colors that make up sunlight have different wavelengths. When sunlight hits water droplets in the atmosphere, each water droplet acts like a tiny prism, with the sunlight entering, refracting, reflecting off the back surface of the droplet, and finally refracting again as the light leaves the front of the water droplet. The colors are arranged from the shortest wavelengths (violet) to the longest wavelengths (red), and the result, from our point of view, is a rainbow.

So we have two very different explanations for the existence of rainbows. One is based directly on the Bible, a straight-forward story of a wonderful promise from God, and the other is a fairly simple scientific explanation of what happens when light passes through water. And having two such explanations is not terribly surprising. Because the physical properties of light and the effect of prisms were not well known until the early eighteenth century, it would be surprising to see them mentioned in a document written more than 2,000 years ago.

But is there a problem here? That is, do the two explanations mean that one of them is necessarily "wrong"? Do we have to decide which one we want to believe, completely rejecting the other? Or is it possible that both views can be considered "correct" for what they are? In other words, can a natural phenomenon be given a scientific explanation and yet be mentioned in the Bible as (and believed by Christians to be) a sign from, and a creation of, God? This apparent dilemma, reduced to a very simple example here, is the crux of this book and the central problem underlying the evolution/creationism issue.

A closely related question is: How much do you want to know about the world around you? Strange as it may seem, for some people, learning that the colors of the rainbow are generated by sunlight passing through drops of water in our atmosphere is unwelcome news, since to them it seems to counter what is so clearly stated in Genesis. I feel almost unspeakably sad for them. Not only are they limiting their understanding of the natural world by shutting out what we have learned about it, they are also limiting the richness of the Bible, more or less condemning it to being a dated document that is relevant only for a time and a place long ago. Similarly, they are limiting God, assuming that God could not employ something as complex as physics and prisms in the natural world. But I do understand their position, possibly because I have encountered it so often. It can be stated as follows:

> I liked the earlier story about rainbows; it is simple and clear, and I am reassured by God's presence in it. The "light and water" explanation takes God out of the picture and replaces him with a cold, unfeeling prism. And prisms are not mentioned anywhere in the Bible. Perhaps I should not believe it.

An oversimplification, perhaps, but it is not that uncommon a position. And whereas it might be hard to find someone with this view as concerns rainbows and prisms, it is quite easy to find people with similar views about other explanations of natural phenomena, such as the age of the earth and the relationships among plants and animals. An alternative viewpoint, and one that is held by the majority of Christians, might be stated as follows:

> The God that I worship seems entirely capable of understanding and employing the physics of light and water; this does not bother me at all. The larger, more important message in Genesis 9 has to do with God's forgiveness, grace, and promise. The scientific explanation of how a rainbow is produced is neither contradictory to, nor really necessary for, the essence and the message of this story.

For Christians, there are several other important questions that need to be asked when comparing the literal wording of biblical passages to scientific explanations of natural phenomena. One such question, still using rainbows and prisms as our example, is: Does knowing that sunlight splits into different wavelengths when it passes through a prism somehow lessen your appreciation of nature, or of God? Some people are far more comfortable simply not knowing, if that knowledge causes them to feel separated from God or at odds with their interpretation of the Bible. We have to respect that, and we must respect them. But for others, the appreciation of the mechanics that underlie the wonder and majesty of a rainbow, and the possibility that it is still God's handiwork regardless of how it is made, is heightened by the knowledge of how it works, not cheapened by it.[2]

# 3

# The Flat Earth Society

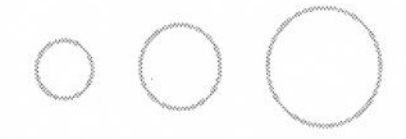

*You set the earth on its foundations, so that it shall never be shaken.*

PSALM 104:5

*The world is round, and the place which may seem like the end may also be only the beginning.*

IVY BAKER PRIEST

It might seem strange that I now switch to a discussion of whether the earth is flat after stating that this book deals primarily with science, faith, evolution, and creationism. But I want to make a slight diversion here in order to introduce a problem arising from "biblical inerrancy," the belief that the Bible is without error with regard to history or science, and "biblical literalism," the belief that every word of the Bible is meant to be literally true. The themes of biblical literalism and inerrancy will also be treated in chapter 4, using a different type of example, because they play a large role in discussions about the interaction of science and Christianity.

Let me begin by saying that the myth that there was widespread belief in a "flat earth" throughout the Middle Ages is exactly that: a myth. People of many different religious beliefs have speculated about the shape of the planet for more than 4,000 years.[1] It is nearly impossible not to notice the slight curvature of the earth when you are out at sea, and people have been going to sea since earliest recorded history. It's clear that most people who thought about the earth's shape, Christians included, have realized that we live on a globe for a long time.

What I'm talking about here is not the Dark Ages but a far more recent phenomenon that included outspoken leaders, beliefs, writings, lawsuits, and a society that was alive and well throughout the twentieth century. The details of this fascinating period in history are given in Christine Garwood's book *Flat Earth: The History of an Infamous Idea* (2007).

When I first heard about the Flat Earth Society I thought it was probably just a bunch of people having fun and kidding around. It's not. Or at least it wasn't at the time. The Flat Earth Society was headquartered just east of Lancaster, California, not far from where I am writing this. And their members claimed to be quite serious about their belief that the world was flat. They pointed to different types of evidence—easy to do in an area as flat as Lancaster and the surrounding desert—that supported their view, and they believed that teaching about a round earth was some kind of conspiracy, as was, in their minds, the space shuttle and NASA in general. What's more, at least some of them were convinced that their belief in a flat earth was completely supported by an inerrant reading of the Bible.[2] Others were far more humble, believing the earth was flat because of their sincere but unsophisticated religious faith and their lack of opportunity for a proper education.

There are many Bible verses that flat-earth believers have pointed to in order to defend their position. Perhaps the best known of these is Revelation 20:7-8, where the earth is depicted as having four corners.

Some early Christian literalists even compiled lists of such verses that numbered into the hundreds.[3] There are even more verses that would seem to support the idea that the earth is fixed (immovable).[4] Perhaps not surprisingly, arguments for a "fixed" earth are also put forth from time to time by flat-earth believers.

You might think that I am describing the Flat Earth Society in order to ridicule its beliefs and its members, but I am not. They have the right to believe what they wish, and they have the right to try as hard as they can to convince me (and others) that they are correct; it's part of what makes an open society work well. And again, I most definitely am not saying that all Christians, as a group, ever believed that the world was flat. But some Christians believed in a flat earth (as apparently some still do) based on their interpretation of parts of the Bible, because they thought the Bible instructed them to believe it, as noted above—just as some have argued against the earth circling the sun, and some have argued about the age of the earth, and some have argued against biological evolution.

Why am I telling you all this? Few of us believe today that the earth is flat. So what's the big deal? Does this mean that the Bible is therefore wrong, not only once but possibly in hundreds of places? Remember the prism and rainbow example, for the same explanation holds here. Rather than being "wrong," perhaps the Bible is telling us something more profound, something deeper, something that goes well beyond the superficial story that also includes, almost anecdotally, phrases that refer to the shape or immovability of the planet. Perhaps what's wrong is the insistence upon a purely literal or inerrant interpretation of passages, which, as we've just seen, can lead to trouble. Such an interpretation often forces us to make a choice, and in this case a literal reading forced at least some people to choose a fixed and flat earth.[5]

What is far more troubling from a Christian perspective is that a purely literal translation of the Bible may also cause us to miss underlying, profound messages while we trip over details that are unimportant. This is not a new problem, of course. Jesus often taught important lessons by using parables, which people would frequently misunderstand by interpreting them literally. ("With many such parables he spoke the word to them, as they were able to hear it; he did not speak to them except in parables, but he explained everything in private to his disciples" [Mark 4:33]). One of Christianity's greatest theologians, St. Augustine, warned Christians against interpreting the Bible literally, as have many other respected theologians throughout history.[6]

One more important question should be asked before we move on: Why don't we teach both a flat-earth/fixed-earth view and a round-earth view of the world in our schools and let the students decide for themselves which idea has the strongest supportive evidence? This is what the advocates of "intelligent design" are proposing when they ask schools to "teach the controversy" (see chapter 9). What could be more fair than that? If we honor free speech, what's wrong with giving students both views of the earth, round and flat, and letting them decide which is correct?

There are several reasons why teaching both views of the earth would be wrong—very wrong. First, everything we have learned in the past 200 years or so strongly supports the view that the earth is spherical, or at least not flat (our current scientific understanding is that it actually bulges outward slightly around the equator as opposed to being perfectly round). According to the best methods and most accurate empirical knowledge available to us—which is to say, according to science—the earth is more or less spherical. Thanks to science, we can even see Earth from space now, something earlier cultures could not do. Thus, teaching that the earth might be round, but also might be flat, would be horribly misleading, especially if the teacher did not have the time to allow students to test this concept directly themselves. Teaching "both sides" would set up students for ridicule and confusion later in life. Second, although flat-earth believers have sometimes pointed to "evidence" that the earth is flat (such as the surface of large bodies of water), most of the arguments for believing in a flat earth have been based on a set of religious writings rather than on science (see Garwood's book). If we were to allow teaching about a flat or fixed earth for that reason, then should we not also teach about other religious views of the shape of the earth? And of course we could do this only in a comparative religion class, not in the science classroom, since such views are so clearly not science.

All of this is a longer-than-necessary way of saying that (1) the earth is not flat; (2) some people in the past believed that it was, and some of them based their belief on a literal or inerrant interpretation of the Bible; (3) they got over it, and Christianity survived just fine, since it turned out that a "round-earth view" was not anti-Christian or anti-religious in any way; and (4) we should not teach, in a public school setting, a clearly religion-based view of the world as a valid alternative to our scientific understanding of the natural world.

# 4

# Of Serpents and Certainty

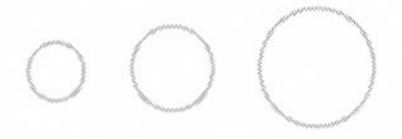

*They will pick up snakes in their hands, and if they drink any deadly thing, it will not hurt them.*

MARK 16:18

*We encounter God in the ordinariness of life, not in the search for spiritual highs and extraordinary, mystical experiences, but in our simple presence in life.*

BRENNAN MANNING

In the small coal mining community of Jolo, West Virginia, in the beautiful Blue Ridge Mountains close to where I grew up, there is a church of the "Signs Following" movement. Members of this movement take literally the passage in Mark where Jesus states:

> And these signs will accompany those who believe: by using my name they will cast out demons; they will speak in new tongues; they will pick up snakes in their hands, and if they drink any deadly thing, it will not hurt them; they will lay their hands on the sick and they will recover. (Mark 16:17–18)

Note that there is not much ambiguity in these words. It seems to say pretty clearly that, if you believe, you will pick up snakes (serpents in some translations) and drink deadly things and suffer no adverse consequences. So as evidence of the strength of their faith, members of the Jolo congregation go into the surrounding hills and gather venomous snakes, usually copperheads and eastern timber rattlesnakes, and bring them back to the church, where they are placed in a wooden box. Congregants who feel moved to do so will come forward and pick up a snake in their hands, sometimes letting it twine around their arms and neck, dancing with the snakes in what has to be one of the most unusual displays of the Christian faith that I have ever seen. Not surprisingly, people are bitten from time to time, and quite a few have died over the years. When handled, snakes will occasionally bite, and when the snake is venomous the results can be fatal. Church members also will occasionally drink strychnine, their poison of choice, as their response to the part of the verse that states "if they drink any deadly thing, it shall not hurt them."

This is admittedly a very small branch on the Christian tree, and snake handling is not a widespread phenomenon. According to Thomas Burton and David Kimbrough, who have both written about the subject,[1] the practice of snake handling originated in Tennessee (started by a man named George Hensley who died in 1955 of—you guessed it—a snake bite) and then spread to the Appalachian areas of Kentucky, North Carolina, Virginia, and Georgia (and also to Alabama and parts of Ohio, Indiana, and Florida). But it is now a rare (and in most states illegal) practice. I mention it here only because of the strength of conviction that these people have. They are certain. To them, the passage in Mark means what it says it means, and if you have some other interpretation then you must be picking and choosing what you want to believe of God's written word. In their minds they are righteous true believers.[2]

I have to admit that, although I strongly doubt that the Signs Following churchgoers have it right, I cannot speak with any authority on what this passage in Mark means either. And that's not too surprising. As I said earlier, I am not a Bible scholar, so for me to pretend that I understand the full meaning of any biblical passage would be both arrogant and wrong. What I can say is that it seems fairly clear to me that something has been misinterpreted or misunderstood here. Indeed, this "longer ending" of Mark is not found in all versions of the Gospel, and many scholars believe it was added later.[3]

Whatever these lines in the Gospel of Mark were intended to imply, and whenever they might have been added, you can pretty much bet they were not intended to teach us that life is cheap. Taking a foolish risk for no apparent reason other than to demonstrate your faith is not a smart move, and exposing yourself to a potentially fatal snake bite or to strychnine is not consistent with any of Christ's other teachings.[4]

There is a theme connecting the Flat Earth and Jolo examples above that by now is probably pretty obvious to you. Both groups of believers claim (or claimed) to be basing their views, or at least some of their views, on the Bible. Both groups claim to be reading the Bible literally or inerrantly; indeed they see no other way to interpret it. To them, any other interpretation amounts to selecting what parts of the Bible you will believe, or reinventing the Christian faith so that it pleases you, rather than strictly adhering to the written word of God. To them, following the Bible completely, literally, is an all-or-nothing thing. Both groups claim to be devout Christians because they follow the word of God in their unwavering adherence to Holy Scripture. What, if anything, are they doing wrong?

# 5

# The Nature of Science

*An intelligent mind acquires knowledge,*
*and the ear of the wise seeks knowledge.*

PROVERBS 18:15

*The whole of science is nothing more than a refinement*
*of everyday thinking.*

ALBERT EINSTEIN

The Flat Earth Society and the Snake Handlers of Jolo represent two distinct but related problems. They are similar in that both groups have claimed that their beliefs are biblically based; they feel that they are following the scriptures as closely as they know how. But they are different because the problem provided by the Jolo snake dancers is not easily dealt with by science. In reality it has essentially nothing to do with science. It is, instead, a problem of biblical interpretation, historical scholarship, and literary analysis. The flat-earth question, however, is clearly and easily dealt with by the methods of science. Simply put: all measurements and observations we have made over the past couple of centuries point to a spherical earth.

So what is science? My dictionary defines science as "the observation, identification, description, experimental investigation, and theoretical explanation of phenomena." You've probably heard that before. And I will not bore you by going into detail about the "scientific method," the process of making observations, drawing a hypothesis, testing it, and moving onward, although it really is important. In fact, it's critical, since it draws such a sharp line between what we know and what we *think* we *might* know about nature.[1] The basic idea is that by applying the methods of science to the world around us we can develop better and better explanations for the causes of natural phenomena as time goes by.[2]

Progress made this way is sometimes painfully slow in coming. Observe some phenomenon, make your best hypothesis, test it, and move on. Science progresses little by little most of the time, with only occasional great leaps forward in our understanding of the natural world. But the results can be stunning, and I shudder to think where humanity would be without this gift. Without science we would not have penicillin, aspirin, the space shuttle, electric lights, cell phones, GPS technology, microscopes, open heart surgery, or the laptop computer I'm using to write this. Because of science, childbirth is now a relatively safe and joyous experience for most women rather than the life-threatening process it was a few generations ago. In many areas of the world, we no longer have to worry about losing our children to common childhood illnesses. If today's medicine had been around in the 1950s, my oldest brother, who died shortly after birth from complications arising from preeclampsia, might have lived.

Instant communication among world leaders now offers the potential for a better understanding of global issues, quick responses to worldwide disasters, and solutions for peace. Planets in our solar system (and even in other solar systems) are beginning to yield their secrets as we explore with satellites and rovers. The floor of the ocean, once thought to be lifeless and inaccessible, now provides us with exciting discoveries on a regular basis. And these advances barely scratch the surface of what science has revealed to us. Without science, our understanding of the world around us would be vastly reduced.

Science is repeatable, meaning that if the same methods are applied and the same conditions hold, we should be able to obtain the same results today that we obtained yesterday. That's an important point, and it also forms a clear distinction between scientific and non-scientific approaches (such as astrology) to understanding the world around us. This repeatability also allows us to make predictions and to test these predictions. Science is also "self-correcting" over the long haul, meaning that we continue to test and retest every idea, every hypothesis, every experimental finding, over and over. That doesn't mean that we have all of the answers in hand now, but it does mean that we get closer to finding out the answers as time goes by.

One of the quotes I used at the start of this chapter, by no less a thinker than Albert Einstein, is in fact a little misleading. Although the methods of science can be characterized as "a refinement of every-day thinking," the results often take us far from what common sense might tell us. Often the findings of science are partly or completely counterintuitive. For example, Christians in the nineteenth and early twentieth centuries who believed strongly in a flat earth based on the Bible (see chapter 3) pointed to the common sense observation that it does not feel to us as though we are living on a round ball that is traveling through space at roughly 67,000 miles (107,000 km) per hour while rotating on its axis at more than 1,000 miles per hour. Science tells us that we are. So science does not always produce the answers that we think it will—it produces, instead, answers that are often surprising. That's part of its strength, part of its appeal. We are never certain beforehand what we will learn.

Science is one of the coolest things that humans have ever developed. It's a process and a way of thinking. It's not perfect, and mistakes are often made (just as they are in any other human endeavor). Science is clearly the best way we have of finding answers about the world around us, or at least certain *kinds* of answers about the world around us, I should say. Science is limited, of course, to addressing only those questions that deal with naturalistic phenomena, the "what, when, and how" questions of the natural world.[3] Science does not try to address the "why" questions of human existence, those vexing, ever-present questions such as "who am I?" and "what am I doing here?" and "what does it all mean?" Nor does it try to address such important human issues as our sense of morality or the value of beauty. The misunderstanding of what science is, and how it operates, is widespread, accounting to a large degree for the misunderstanding about evolution and other sciences, not only among fundamentalist and evangelical[4] Christians but within the general public.

I mentioned that progress sometimes is slow in coming—a result of the painstaking care with which scientific work is supposed to be conducted—but it's also important to realize that the understanding of scientific advances by the non-scientific public can also be quite slow in coming. The Galileo incident is one of the best examples of the lag time between the acquisition of scientific understanding and public acceptance of that understanding. Galileo was arrested in 1633 by Roman Catholic Church officials for expressing views about the earth that they thought were heretical (see chapter 11). He eventually received an admission that he was correct from the Roman Catholic Church . . . but not until 1992.[5] They got over it, Christianity survived after all, and people eventually accepted the fact that the earth was not the center of the universe. It just took time.

In a way, a good team of scientists—say, for example, the excellent group of engineers and physicists working at the Jet Propulsion Lab in Pasadena, California—operates in the same way that a good sports team works: The best players rise to the top and play, regardless of their race or religion. If a coach wants her team to win, she does not discriminate on the basis of race or religious beliefs; she puts her best team in. The same thing is true in science. Any group of serious scientists is (or should be) open to a variety of backgrounds and beliefs. If your goal is to build a new deep-sea submarine, or a space shuttle, or develop a cure for ovarian cancer, or study the history of life on Earth, then you want the best-educated, most gifted scientists in the world on your team, regardless of whether they are Muslims, Hindus, Christians, Buddhists, Jews, or members of any other religious (or non-religious) group. The goal is always to find out, to the best of our abilities, how the world around us works.

There is nothing evil about science, nothing anti-religious, nothing anti-Christian. This perception has arisen because of the occasional conflicts between the findings of science and a literal or inerrant interpretation of the Bible. There also is no intentional bias or animosity among scientists as a group toward any religion or religious groups. Matters of faith are, quite simply, out of the realm of what science can address. Moreover, active scientists usually don't have the time, or any incentive, to wage some sort of attack on Christianity or organized religion; they are usually far more concerned with learning as much as possible about the world around them in the limited time we are given on Earth.

Individual scientists, of course, do sometimes state that, based on science, there is no God, or there is no basis for religion. The prolific writer and evolutionary biologist Richard Dawkins is probably the best known of these, but he is certainly not alone in making unequivocal statements along these lines. To argue that faith of any kind is bogus, archaic, or ultimately damaging, as he does in his 2006 book *The God Delusion*, is reaching beyond the bounds of what science can rightfully address. Many other writers have similarly attacked religion in general, with titles that seem intentionally penned to be divisive, as in Christopher Hitchens's 2007 book *God Is Not Great: How Religion Poisons Everything* and the rather incredibly titled *God: The Failed Hypothesis—How Science Shows That God Does Not Exist,* by physicist Victor J. Stenger.

Religion certainly is not immune to criticism, and all of these authors make some valid points and raise legitimate concerns. The history of Christianity is full of horrendous mistakes and offenses committed in the name of God, and we would do well to study these and not sweep them under the rug. These writers are also free to profess their complete lack of any personal faith or belief in God. But they cannot do so based on science, a creative endeavor specifically designed to probe the natural rather than the supernatural. The scientific method is designed and used to test ideas, not to make subjective judgments about value, morality, or faith. Because of the limitations of what science was created to address, writers who base arguments against faith on science have, I would argue, in this instance overstepped their bounds.

# 6

# What Does "Theory" Mean?

*In theory, there is no difference between theory and practice.*
*But in practice, there is.*

YOGI BERRA

*There is nothing more deceptive than an obvious fact.*

ARTHUR CONAN DOYLE,
*THE ADVENTURES OF SHERLOCK HOLMES*

I should discuss briefly the confusing term "theory," used in most of today's conversations to mean a good guess, a hunch, or an idea, where your theory about why Mrs. Johnson is having a bad hair day is just as good as mine. In science, the word has a very different meaning: in science, "theory" indicates a well-supported body of explanatory statements that includes facts, laws, rules, tested hypotheses, and everything else we know about it. Scientific theories are large, overarching frameworks into which the observed data fit. Additionally, in science a "current theory" is one that has no equally acceptable alternative theory and that has survived attempts at falsification.

Accumulating more information does not turn "theory" into "fact." Indeed, nothing in science is really ever proven (apart from the special use of the word "proof" in math). We just continue to add more knowledge and gain more understanding as time goes by.[1]

Some things become so well known after many years of study that it seems odd to continue to refer to them as theories, because of the common use of that word. Examples include gravitational theory, still often referred to as such despite the fact that gravity itself is generally well accepted. Gravity is the phenomenon, whereas gravitational theory is the explanation of it. Other examples include germ theory and cell theory, by which we mean essentially everything we know about germs (the body of evidence and explanations telling us that microorganisms are responsible for many diseases) and everything we know about cells (the body of evidence and explanations telling us that cells are a fundamental component of living organisms); the atomic theory in chemistry is another example. Thus, comments such as "it's only a theory," meant to imply considerable room for doubt, are not applicable to evolution any more than they are to gravity or to our current understanding of our solar system. The heliocentric theory of the solar system is still called a theory, but that does not mean that it's just a guess or a hunch that the earth revolves around the sun. One would be hard pressed to find a student or a teacher today who doubts that the earth orbits the sun since heliocentrism is "just a theory."[2]

Words, as insufficient as they are, are almost all that we have to describe the world around us to others. So we need to be clear about our use of terms that are employed differently in science and in common speech. If gravity (or rather gravitational theory) is, as noted above, rightly considered a theory, then what is a fact, and what are hypotheses?

**Fact**. It seems like everyone knows what a fact is. And yet, the word has a couple of meanings that are not quite the same, and the word "fact" is thrown about incautiously by many people, including many scientists. The first, more restrictive, and more correct definition of the word is an object or event "having real, demonstrable existence . . . the quality of being real or actual."[3] That seems simple enough. But it means that gravitational theory, as well known and accepted as it is, technically is not a fact. Rather, the *fact* is that the apple fell from the tree, or the dropped coffee mug crashed to the floor (more precisely—they both fell toward the center of the planet). Gravitational theory is the overarching theory that *explains* and *includes* the facts. This seems quite odd to us, to think that gravity is not a fact in this sense of the word, when we certainly do not doubt it. Exactly the same is true for evolution. In this first definition of "fact," evolution also does not fit; it is instead the causal theory that *includes* and *explains* the facts.

But "fact" has also come to mean something that has been observed so often or is so well tested that it is no longer doubted. Even the National Academy of Sciences (NAS), in their very helpful booklet *Teaching about Evolution and the Nature of Science*,[4] used it in this way, defining it in the context of science as "an observation that has been repeatedly confirmed." I sort of wish they (and many others) had not done so, since repeatedly observing something does not really make it a fact, and because it causes confusion with the first definition given above. For example, I can observe the sun rising in the east morning after morning, and it certainly appeared to do that in the eyes of people of ancient times, but of course it does not really rise at all, no matter how often I observe it.

You and I know, of course, what these people mean when they use words this way—they mean that we no longer doubt that gravity occurs, and so it is a "fact" in that sense. It is this second sense of the word that scientists and others use when they refer to gravity as being a fact, or when they make statements such as "evolution is as much a fact as anything we know in science."[5] Scientists certainly do not doubt the attraction of bodies of mass to one another (gravity) or the change in populations of organisms over successive generations (evolution), and so we sometimes refer to them as facts. But this dual use of the word causes confusion, as both are more properly called theories. It is also important to note that a fact in this second sense of the word is often the result of indirect, rather than direct, observations. We often use inference (deriving a logical consequence from a set of premises or data) to make observations about things that cannot be directly observed or measured, such as the earth orbiting the sun, the function of mitochondria, the age of the earth, the genetic basis for cancer, the fact that *Tyrannosaurus rex* was a meat eater, etc.

**Hypothesis.** A hypothesis is merely a suggested explanation for something. The "something" might be an object or a series of events that seem to be related. As commonly used in everyday language, a hypothesis is usually considered as being in need of further evaluation, much like a good first guess. In science, there is the additional restriction that a hypothesis has to be testable, otherwise the suggestion does not tell us anything. I could suggest, for example, that my cat exhibits erratic behavior because he was abducted by aliens in one of his previous lives. That's certainly one hypothesis. And who knows, it might even be right. But if I cannot come up with a way to test it, then it's of no use as a scientific hypothesis. A better hypothesis (better because it is testable, not because it is necessarily correct) might be that he is allergic to tuna fish. The words "hypothesis" and "theory" are often used synonymously in everyday language, but in science they are different things.

**Law.** A scientific law is a statement that describes a particular set of behaviors or associations in the natural world. Often it describes a connection between physical quantities. Because the term is so often associated with physics, it is sometimes used synonymously with "physical law." Such laws refer to principles that are thought to be universal in nature, and sometimes they are called "principles." They often contain an analytical statement and/or a "constant" (number) used in an equation. Examples are Boyle's gas law (the inverse relationship between pressure and volume of a gas at a constant temperature), the law of conservation of matter in chemistry (which states that in a closed system, matter cannot be created or destroyed, just rearranged), and some of Newton's classic laws of physics (e.g., a body at rest tends to stay at rest; force equals mass times acceleration; and for every action there is an equal and opposite reaction). The biological sciences also have scientific laws, such as Mendel's laws of inheritance (concerning the segregation and independent assortment of alleles during meiosis) and the Hardy-Weinberg law (often called the Hardy-Weinberg principle) stating that allele and gene frequencies in a population remain constant from generation to generation unless something is introduced to disturb the equilibrium. Laws can be shown later to be false, despite their name.

**Rule**. The word "rule" when used by biologists refers to apparent patterns that seem to be widespread in nature; they are usually named for the person who first noticed or described the pattern. Examples include Bergmann's rule (which says that, within a species, animals tend to have larger body sizes in colder climates), Allen's rule (populations in colder climates tend to have shorter limbs), and Cope's rule (the average size of individuals in successive populations tends to get larger over geologic time). Rule, when used in this manner, means "generally," not "invariably." Despite all these patterns being referred to as rules, there can be (and usually are) exceptions to them, they are not necessarily accepted by all scientists, and they are subject to testing and discarding when found to be untrue.

**Theory**. As I noted at the start of this chapter, in science a theory is used to refer to the entire body of interlocking and well-supported explanatory statements about a given topic, as when we speak of gravitational theory, germ theory, or evolutionary theory. The NAS uses the following definition: "A well-substantiated explanation of some aspect of the natural world that can incorporate facts, laws, inferences, and tested hypotheses." Use of the term "theory" does not imply doubt about gravity or germs or evolution; it's just how we refer to the body of knowledge (including facts, hypotheses, and all sorts of evidence, as noted above). In stark contrast to everyday use, where "theory" indicates the absence of supporting information, in science it's just the opposite: in science, theories are the ultimate goal.

I admit it gets messy, because the word "theory" can also be used to refer to the various proposed *mechanisms* that underlie gravity, evolution, and other phenomena. And these mechanisms *are* still very much under study. Gravity is a good example. Although we no longer doubt gravity, we still do not know exactly how it works. Einstein's theory (general relativity) displaced Newton's ideas[6] in the twentieth century, but it is still open to revision should a better idea come along.[7] In the same sense, natural selection was one *mechanism* that Darwin proposed to explain evolution. Prior to Darwin's work, other workers had come up with different mechanisms to explain the obvious changes in life over time. One of the competing ideas was the inheritance of acquired characteristics (the idea that environmentally caused changes affecting an animal during its lifetime, such as losing a tail or stretching a neck, could be passed on to its offspring), championed by Jean-Baptiste Lamarck and others. That idea was replaced for the most part by Darwin's natural selection. Used in this way, "theory" is a testable explanation, much like a hypothesis. As a result, sometimes you will read that there is a distinction between the "fact" of evolution and the "theory" of natural selection. Today we no longer doubt whether evolution has occurred, but the mechanisms for it are often debated. This honest, intellectual debate is sometimes (erroneously) seen as a debate over whether evolution has, in fact, occurred. It isn't.

It's unfortunate that so many of these terms can have more than one meaning and are used inconsistently by both scientists and nonscientists. Perhaps a phrase such as "Grand Unifying Theory" or a word like "paradigm" should be used to set off the first usage (the overarching body of explanatory statements) from the second (the everyday use) of the word "theory," but that has not happened. It is partly for this reason that we sometimes see statements such as "evolution is both theory and fact" in biological literature. That statement is meant to imply that we no longer doubt that evolution has occurred and continues to occur (the second definition of "fact," above) and also that "evolutionary theory" is a well-tested framework into which all of these observations fit (the first definition of "theory," above); it can also mean that the mechanisms (such as natural selection and genetic drift) continue to be tested and studied. It does not mean that it is "a possibility and nothing more."

# 7

# What Is Evolution?

*The fruit of the righteous is a tree of life.*

PROVERBS 11:30

*Everything is simpler than you think and at the same time more complex than you imagine.*

JOHANN WOLFGANG VON GOETHE

Evolutionary biology is the study of the history of life on Earth and of the processes that have led to life's diversity. It's one of the biological sciences. And yet it's also more than just "one of the biological sciences" in that it gives us a broad, general framework that *unites* the biological sciences. Fields of study as diverse as physiology, behavior, ecology, genetics, biochemistry, medicine, and more are all supported by, and linked by, an understanding of evolutionary theory.[1] Furthermore, a better understanding of evolutionary theory is critical to our survival. If ever we are to solve such problems as the spread of HIV, SARS, avian and swine (H1N1) flu, and other rapidly evolving infectious diseases, if ever we are to gain the upper hand on genetically inherited disorders, it will be because we have gained an understanding of how such maladies arose and have changed, and are continuing to change, over time.

The main concepts of evolution can be summarized fairly simply. The following description (with only slight modifications) comes from the "Evolution 101" section of the web site "Understanding Evolution," a joint project of the University of California Museum of Paleontology and the National Center for Science Education:

> Biological evolution, simply put, is descent with modification. This definition encompasses small-scale evolution (changes in gene frequency in a population from one generation to the next) and large-scale evolution (the descent of different species from a common ancestor over many generations). Yet biological evolution is not simply a matter of change over time. Lots of things change over time: trees lose their leaves, and mountain ranges rise and erode. But these things aren't examples of biological evolution because they don't involve descent through genetic inheritance.
>
> The central idea of biological evolution is that all life on Earth shares a common ancestor, just as you and your cousins share a common grandmother. Through the process of descent with modification, life on Earth gave rise to the fantastic diversity that we see documented in the fossil record and around us today. Evolution means that we're all distant cousins: humans and oak trees, hummingbirds and whales.

Thus, the central idea of evolution is that we share common ancestors. We are descended from these ancestors, but modifications have arisen along the way. "Descent with modification," perhaps the simplest and most often used "thumbnail" definition of evolution, simply means that over time, from one generation to the following generations, slight changes accrue in a population. We descend from our grandparents, as they did from theirs. But we are not identical to them. And it is the same in all organisms.

One of the cornerstones of evolutionary theory is natural selection, an idea that can be expressed in five simple statements:

1. There is a lot of variation in nature. Individual organisms in a population—whether trees, rabbits, or bacteria—are not identical.
2. In nature, organisms tend to produce far more offspring than can survive, something particularly obvious in plants and insects. Some produce hundreds of eggs or seeds; some produce many millions.
3. Not all of these eggs and seeds will make it. Some will survive and some will not.
4. The individuals that possess traits that are most useful in the environment in which they find themselves are the ones most likely to survive and (more importantly) pass on their genes to the next generation.
5. Over time,[2] the population changes as some traits are preserved and others are eliminated, such that future populations will differ from earlier ones.

Humans knew for hundreds of years before any of Darwin's writings that plants and animals could be intentionally bred so that certain traits were preserved, and others eliminated, in the next generation. We do this routinely today (as we did way back then) with domestic farm animals, pets, crops, flowers, microbes, and many other organisms to "artificially select" the traits we want to see in the next generation. It's how breeds of dogs and cats are created (and maintained), how we make cattle that produce more milk or more beef, how we create fruits without seeds.

In a sense, the discovery of "natural selection" was simply the recognition that this kind of selecting also occurs in nature. Because humans are not doing the selecting in nature, we usually call this "natural selection" as opposed to human-caused "artificial selection." Darwin is often credited with "discovering" evolution, but this is not true. Many people of his day and well before then understood that biological change, including the formation of new species, occurs over time.[3] Darwin's main contribution was to supply some of the details that support the rather simple idea of natural selection—which is one mechanism by which evolution operates.

It would be wrong of me to imply that evolution is in any way simple. What I have described above concerns mostly natural selection, but other factors are involved. The patterns and mechanisms of evolutionary change, including natural selection, mutation, migration, gene flow, and genetic drift (a change in gene frequencies of subsequent populations that is not due to adaptation), can be very complex. University of Chicago biologist Jerry Coyne, in his recent book on evolution,[4] supplies a single sentence that captures the essence of evolution and then follows it by noting the six major components of evolutionary theory: evolution, gradualism, speciation, common ancestry, natural selection, and non-selective mechanisms of evolutionary change. This is why evolution is most often taught at the university level as a standalone, full-semester class that demands a working knowledge of biology, math, and genetics, and why there are so many scientific journals and books devoted to this field of study. There is still much that we do not know about evolution (as is true of all sciences). Biologists are working hard to figure out how it happens, and there is much research and many healthy debates concerning the mechanisms involved. But as to whether evolution has occurred, among scientists there is no debate and no controversy.[5]

The evidence for biological evolution is clear, unambiguous, compelling, and overwhelming, easily as strong as the evidence for a spherical earth. All this evidence supports the central concept of evolutionary theory: that life on Earth has evolved and diversified into many forms and that species share common ancestors. Although this book is not the place to try to review all of the evidence (in keeping with my desire to keep it short), the evidence includes findings in the fields of geology, paleontology, astrophysics, genetics, molecular biology, developmental biology, and more.[6]

Furthermore, this enormous amount of evidence has not been "manufactured" by science or by scientists. Just as a prism does not create the colors within sunlight but only reveals them, so science merely reveals what is already present. The evidence is part of the natural world around you. For people of faith, this is seen as part of God's world. And if you decide to go look for yourself, you will find it.

Hundreds of religious leaders, Protestant and Catholic, including Billy Graham, Pope Pius XII, and Pope John Paul II, have indicated their acceptance of evolution as a science that is not in conflict with faith.[7] Courses in evolutionary biology are taught at nearly every major university in the United States and in most other countries. The greatest minds and universities on Earth have seen the importance of evolutionary biology as a major field of science.[8] I am not saying that you should always vote with the majority—the majority has been wrong often in the past. What I am saying is that in this case, you might want to ask yourself why the majority—not only the overwhelming majority of the world's scientists, but also so many of the world's religious leaders—believe what they do.

And yet, because evolution appears to be in conflict with a literal or inerrant reading of the book of Genesis, many Christians are uncomfortable with it. What's important to ask yourself is why this is so, why many people are trying to argue *against* evolution. It's worth noting that the arguments against evolution or against science are almost never based on logic or evidence. The arguments exist because these people feel that their religion is threatened. That is, if religion were completely removed from the picture, I think most of the arguments would cease. What I hope I have shown you is that these advances in knowledge—the prism, the shape of the earth, and evolution—do not in any way threaten Christianity or any other religion. They are what they are, revelations of the way the world works that we have discovered. Revelations, in other words, of God's world.

Theologian and writer Nancey Murphy, of Fuller Theological Seminary, reflects on the damage that can be caused when evolution is taught as something that is contradictory to one's faith:

> When I first discovered that there are still Christians who reject evolutionary theory (having grown up in the Catholic school system, I did not encounter this as a child), I thought of it as a harmless expression of ignorance. More recently, though, I've come to see it as tragic. Vast numbers of young people are taught that evolution and Christianity can't both be true. They get a good science education in college, recognize the truth of the evolutionary picture, and then believe that they have to reject their faith.[9]

Evolution is the central organizing framework of biology. Like the prism in our earlier example, it is neither contradictory to, nor really necessary for, the enduring moral and ethical lessons of the Bible. Evolution is no more controversial in scientific circles than gravity, the water cycle, or electricity. And Christianity, which did, after all, survive the startling (and at the time heretical) revelation that the earth circles the sun rather than vice versa, will also survive the revelation that biological evolution is a fact of life. It might take time, as it did with the Copernican revolution, for persons of faith to realize that they, and their faith, will survive. But it will happen.

# 8

# What Is Creationism?

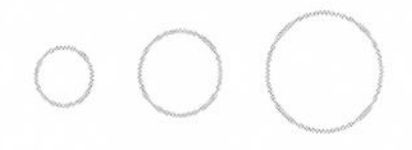

*Have you not known? Have you not heard? The Lord is the everlasting God, the Creator of the ends of the earth.*

ISAIAH 40:28

*Biology, chemistry, astronomy, physics, and geology are only at odds with the Bible when the Bible is expected to speak with authority in the language of these topics, and its writings to confirm the discoveries and postulates of these sciences. If this test of science is applied to scripture, scripture will always fail. But having said that, we have said really nothing at all, for scripture does not pretend to be science any more than science pretends to be scripture.*

PETER GOMES, *THE GOOD BOOK*

At its simplest, creationism is the view that the earth and everything in it was created by God and that nothing much has changed in the meantime. Many religions have creation stories, but most often when someone mentions creationism they are referring to the creationism of Christianity, based primarily on Genesis 1 and 2. The details vary according to which creationist is speaking or writing, and there is a surprisingly wide range of stances taken by creationist writers. Some believe in a very young earth (6,000 to 10,000 years old). Some still feel that the world is geocentric (that is, they believe the earth stands still while all else revolves around it).

We saw earlier that there have been some creationists who believed that the earth was flat. Others are more progressive, acknowledging that the planet really is as old as all the evidence indicates and that it orbits the sun, but who believe inerrantly or literally in other stories in Genesis, including the worldwide flood and the Genesis account of the creation of man and woman.[1] The idea of evolutionary theory is particularly troubling to most creationists, because it seems to counter their literal or inerrant reading of the Genesis account of the creation of plants and animals in six days.

Because of their belief in biblical inerrancy, creationists tie themselves into knots of reasoning when they try to reconcile biblical passages with what we have learned about the world around us over the past 200 years or so. Some creationists have written that the light from distant stars (which would contradict their view of a very young earth) was created "en route" to appear as if it came from stars millions of miles away (and thus millions of years ago). Some have written that fossils were made to look old and then intentionally planted to make the earth appear old, implying that God must be some kind of trickster or master of deception. A more common explanation among creationists is that fossils are genuine but are the result of Noah's flood as described in Genesis. To support the flood hypothesis, some have written that the worldwide distribution and obvious sequence of animals in the fossil record can be explained because the more "advanced" species could run faster to escape the rising flood waters, and this is why they are found in the upper layers of the geologic record, "above" the more primitive animals that could not escape in this manner. If we believe this, we also must assume that deciduous trees, which are found only in higher (more-recent) layers, must have run faster than pines and other coniferous trees, which are found only in earlier (lower) fossil layers. Fishes, too, present a problem to this line of reasoning, as it is unclear why the flood waters would have wiped out only some. Some of the "young-earth creationists" have argued that dinosaurs (sometimes called "creation lizards" to imply that they were created at the same time everything else was) walked with humans (they would have to have done so if the earth is only 6,000 years old). Because of this belief, there are a number of roadside creationist "museums" and playgrounds in the United States showing dinosaurs coexisting peacefully with humans.[2] Some creationists have written that the tiny forearms of *Tyrannosaurus rex* are evidence that it was a fruit-eater (the small, two-fingered claws are claimed to be an adaptation for picking fruit off of trees), in order to explain why it could have coexisted peacefully with humans without, apparently, demanding too much attention. The amazing six-inch carnivorous teeth of *T. rex* are either ignored or, at least on one web site, explained away by noting that some herbivorous animals today have sharp teeth and yet eat only plants . . . And on it goes.[3]

It might be seen as harmless or even humorous except that belief in creationism is so widespread that some state school boards in the United States—Kansas has the dubious honor of being the best known of these, but it is not alone[4]—have tried to open the door to teaching creationism or "creation science" as an alternative to scientific views of the earth by watering down science standards to get evolution off the curriculum. Other school boards, including one in Cobb County, Georgia, have inserted stickers into biology textbooks to cast doubt on evolution, thinking that by so doing they were promoting godliness. In January 2005, a federal court in Georgia ruled (and I would agree) that the Cobb County stickers badly misled students about the nature of science and that the Cobb County Board of Education had "improperly entangled itself with religion by appearing to take a position." The stickers were removed.[5]

When confronted with the obvious problems in the logic of creationism, the creationists turn to arguing against evolution. To do this, they point to honest disagreements and discussions among evolutionary biologists as if that were evidence of a problem (a "theory in crisis"), and they refer to things like gaps in the fossil record, noting that "no one was there to see it," so we cannot know what happened. Another tactic is to stress the godlessness they perceive in the processes of inheritance and selection. Finally, they sometimes argue that evolution itself is a religion, and that people adhering to evolutionary views are essentially people of faith too, so that both views—evolution and creationism—should be taught in schools, or both should be thrown out. These three arguments have been dubbed the "pillars of creationism:"[6] (1) evolution is a "theory in crisis," (2) evolution is atheistic, and (3) it is only fair to "teach both sides."

In fairness, what the creationists fear most is that, if they concede that any part of Genesis is not literally true or inerrant, it would open all of that book to such scrutiny, and then where would it all end? What other parts of their belief would science show to be baseless or misunderstood? If they concede that Genesis was not meant to be historically and scientifically factual, then what does this say about matters of morals and faith? If, for example, there was no historical Adam, what might this say about the Fall, and about Original Sin, and Atonement? Such concepts can be frightening to people who have not read deeply enough to realize that Genesis was never meant to be read or interpreted literally. Their fear is very understandable, and fundamentalists of other religions, and of other ages, have voiced the same concern about "modernism" in general. This fear is also immediately recognizable as the same fear that underlay the reluctance of earlier Christians (and others) to believe that the earth circled the sun.

Although Christianity is the most common religion in the United States, most Christians have no desire to see literal creationism promoted in public schools or even at church. Geneticist Dr. William Thwaites (San Diego State University, now retired), writing about the evolution of creationism on behalf of the National Center for Science Education (NCSE), noted:

> It is a little-known fact that Methodists, Presbyterians, Lutherans, the United Church of Christ, and many other denominations do not believe that Creation occurred liter-

> ally as described in Genesis. In fact, the majority of Christian seminaries do not teach a Biblical literalist creation. In the United States and Canada, one tends to find Biblical literalist beliefs being promoted most strongly in small, independent denominations, where it is not uncommon for the leader to have little or no formal theological training.[7]

And although it is relatively easy to find polls showing whatever result you are looking for, I do think it is instructive to note that a 1999 poll found that only 29% of the U.S. public wanted creationism taught along with, or instead of, evolution.[8]

The mistakes of the creationists are many. The most obvious is their lack of understanding of what constitutes science (a misunderstanding that is not restricted to creationists, of course). But perhaps their largest error is their reliance on physical evidence to support their faith. When you tie your faith to things that can, in fact, be tested, and you find out later that they have been tested and that the results do not support your religious views, you are in for a fall. It is a type of theology that some theologians and philosophers have referred to as the "God of the Gaps" approach, in which God is invoked as an explanation for the areas where human knowledge is incomplete. Ken Miller, in his exposé on the arguments against evolution, expressed his concerns with this approach concisely:

> As a Christian, I find the flow of their logic particularly depressing. Not only does it teach us to fear the acquisition of knowledge, which might at any time *disprove* belief, but it suggests that God dwells only in the shadows of our understanding. I suggest that if God is real, we should be able to find him somewhere else—in the bright light of human knowledge, spiritual *and* scientific.[9]

Miller's words are consistent with those of many great Christian theologians, including Dietrich Bonhoeffer, who wrote from his Nazi prison cell in 1944: "How wrong it is to use God as a stop-gap for the incompleteness of our knowledge . . . We are to find God in what we know, not in what we don't know; God wants us to realize his presence, not in unsolved problems but in those that are solved."[10]

There are many different versions of creationism, and I have not covered all of the creationist views and arguments here.[11] What most creationists have in common is that they believe literally or inerrantly in Genesis (or at least more so than most Christians do), and particularly in the inerrancy of the creation story in Genesis,[12] and that they seem to get quite agitated at any mention of evolution, astrophysics, or any other bodies of evidence that they cannot explain or that cause them to feel that their faith is threatened. In recent years, "creationism," although still alive and well in its original form, has in many ways (particularly in the media and in public school board debates) given way to the phrase and movement known as "intelligent design," which is treated in more detail in the following chapters. As we shall see, intelligent design has been the latest of several attempts to put creationism on a scientific basis by calling it science. It's revealing to see that this is not new, as the following quote concerning "creation science" attests:

> The phenomenon of "creation science" is the potential source of some of the most dangerous societal, educational, and religious problems of our age. Where the distinction between religious history and science becomes confused in the minds of the average citizen, student, or legislator, there is danger to the quality of education generally, and to that of science, philosophy, and theology specifically. A nation whose high school and college science is mediocre can hardly hope to be a world leader in science and engineering research. A nation whose understanding of theology is so meager that it cannot draw a clear distinction between science and religion is educationally impoverished. More-

> over, the confusion on the part of religious fundamentalists and politicians, as has been demonstrated in certain parts of the United States, bodes ill not only for the quality of science education but also for the good name of religion among thinking people. Many scientists and theologians regard this phenomenon as the tip of the iceberg of a fundamentalist religious movement, heralding a possibly well-intentioned but simplistic, anti-intellectual, and potentially dangerous attempted retreat from the complex problems of the twentieth century.

The above passage was written by a Jesuit priest and geologist (Father Jim Skehan) more than twenty-three years ago.[13] It is startling both for its accurate depiction of the "creationism" issue at that time (1986) but also for the prediction contained in the last sentence, some of which, at least, has come to pass. As we shall see in the next chapter, some components of the "creation science" of the 1980s changed into various other forms of creationism, including intelligent design. But it is still around, still recognizable, and still a troubling and potentially dangerous component of life in America.

# 9

# What Is Intelligent Design?

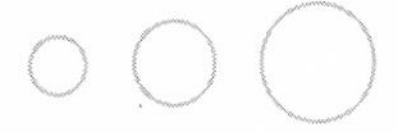

*The greatest enemy of knowledge is not ignorance, it is the illusion of knowledge.*

STEPHEN HAWKING

*The intelligent design movement has the unfortunate effect of promoting the view that science and Christian teaching are incompatible.*

NANCEY MURPHY, FULLER THEOLOGICAL SEMINARY

For some people, the term "intelligent design" seemed like a welcome and recent addition to the conversations about faith and science. For people searching for a way to accept both the findings of modern science and religion, the name sounded like an agreeable meeting place, a gentle compromise where modern science could be accommodated without the loss of God. At its simplest, intelligent design implies that not all of the complexities that we see in nature could have come about by way of biological evolution; in other words, something else must have been responsible. At that level, to a great many people it sounded reasonable. It was not a new idea—

the appearance of design in nature was noted by the ancient Greeks—but it was presented as though it were, and many people were willing to listen.

In truth, the introduction of the term "intelligent design" was little more than a clever way of trying to promote creationism under another name, a name that sounded more scientific and might not be refuted by the American court system. Creationists had lost the battle to ban evolution from the classroom altogether (declared unconstitutional by the Supreme Court in 1968), and they had also been defeated in attempts to require the teaching of creationism as if it were science (court cases of 1982 and 1987).[1] After these defeats, it became apparent to the creationists that they would continue to lose their battles in courts of law (because they were violating the U.S. Constitution's First Amendment's establishment clause concerning the separation of church and state). As a result, some of the more politically astute creationists changed the name of creationism (although creationism itself is still around). For a brief period the name "creation science" was used, in an attempt to give creationism a little more credibility. Because the word "creator" was too obviously religious, some creationists began to use the term intelligent design (ID) instead. It's mostly the same set of arguments but now made by a slightly younger generation of creationists; "creator" simply becomes "designer."[2] In fact, intelligent design is often referred to as IDC for "intelligent design creationism."

But ID is also slightly different from the creationism of the past. Overall, the ID movement is far more sophisticated than the earlier creationist movements. Although officially ID is neutral as concerns the age of the earth, most advocates of ID (though not all of them, interestingly) readily admit its great age. Many of the more vocal proponents are well educated with degrees from impressive institutions (though hardly any of them have degrees in any field directly related to biological evolution). Some (but again, not all) freely admit that evolution has taken place, but they put restrictions on what can, and cannot, be explained by evolution, with the unexplained part being attributed to God (called the "designer" to avoid the legal troubles). The most progressive members of the group will even allow for the existence of "microevolution," usually defined as small changes occurring within a population over short amounts of time—including things we can see in our own lifetime, such as the changes we see in generations of dogs bred and selected for a desired trait—but not "macroevolution," the larger, overarching view of life that involves the formation of new species and new lineages. But their approach is vague and dishonest; they use microevolution to describe variation within "kinds" (because the word occurs in Genesis), without ever describing what a kind is. Regardless of how great the difference is between organisms, they will always say that this difference is just microevolution, simple variation "within a given kind." This allows them to believe (and to say) that there is no evidence of large-scale (macro) evolution at all.

To biologists, microevolutionary processes involve changes occurring within gene pools, small changes occurring within a population. Macroevolutionary processes involve changes occurring across separated gene pools, including the formation of new species and new lineages.[3] It's simply a matter of scale. Speciation is a macroevolutionary event, and so is extinction, since the entire gene pool is wiped out. Thus the creationist or ID argument that allows microevolution but not macroevolution is intellectually empty; it is much like a geologist who believes in some erosion as long as no canyons are formed.

Arguments in support of ID are often clothed in the terminology of science to give them the appearance of credibility (as are some arguments of creation science). These are not simple country folks arguing for a young earth and creating roadside displays where dinosaurs cavort with people. The ID movement is well funded and is for the most part spearheaded by the Center for Science and Culture, which is part of a think tank in Seattle, Washington, called the Discovery Institute. Some of their web sites are beautifully designed, and some of their writing is elegant and forceful. Their arguments on the surface appear to be much better developed than those used by the creationists and are sometimes beautifully articulated, quite different in approach from the young-earth creationists from which they sprang (and from which they now try hard to distance themselves publicly, even though some of their founding and current members are young-earth creationists).

They also have a long-term plan for instituting their views nationally by lobbying high-ranking politicians, a change from the more grass-roots activities of the earlier creationists. Phillip E. Johnson, a retired Berkeley law professor, led the group in carrying out a long-term, three-phase strategy called "The Wedge Strategy" (a title that itself appears to dismiss any attempt at reconciliation or bridging) in an effort to overthrow the teaching of evolution and all secular culture over a period of many years.[4] Perhaps not surprisingly, they have found some senators and representatives, as well as school boards, willing to listen and to advocate for ID, believing that they are representing their constituencies by so doing. And these politicians are also quite skilled, using phrases like "balanced and fair treatments" and "tolerance for diversity" to smoothly argue for inserting this modified brand of creationism into the public schools, and with battle cries like "teach the controversy" that make it sound like they are simply trying to expose students to all possible views on the subject. On the surface, this sounds perfectly reasonable, especially in America, where democracy is so highly valued. And in some ways they are succeeding; an informal survey of the National Science Teachers Association found that nearly one-third of the 1,050 teachers who responded felt that they were being pressured to teach creationism, intelligent design, or other nonscientific alternatives in the science classroom.[5]

Like the creationists, the ID advocates usually argue either that (1) evolution is a flawed theory (a "theory in crisis" again), and that there is some scientific evidence for ID that would allow it to be taught in science classes as an alternative to evolutionary theory, or that (2) evolution is tantamount to atheism (their reasoning here is that teaching atheism would violate the First Amendment) and so we should teach both sides, the "strengths and weaknesses," to be fair. In these ways, their arguments reflect the "pillars of creationism" arguments referred to earlier. But unlike the earlier creationists, who sometimes went so far as to claim that evolution is a religion that should be tossed out of school (if creationism, their religion, is not allowed), ID advocates take a slightly different approach. They do not ask that evolution be thrown out; rather, they argue for what they call an *improved* approach to teaching evolution, one that teaches all the flaws and weaknesses of evolutionary theory, which they say the "dogmatists" do not discuss.

By casting doubt on our modern understanding of evolution and science in general with their "improved" and "more balanced" approach to teaching evolution, they can then offer other, more palatable, theories—namely, intelligent design (creationism). The most recent arguments do not even use the words intelligent design, undoubtedly in light of its 2005 defeat in the Dover, Pennsylvania, federal court. Instead, more recent contributions from the ID advocates argue for "critical analysis"[6] of evolution and for "teaching the controversy," referring to themselves as advocates for "academic freedom." This new strategy, instead of requiring teachers to present criticisms of evolution, simply allows them to do so. The desired result would be about the same; it would eliminate any effective checks on antievolution activities. Interestingly, and predictably, the most recent book on the topic offered by the Discovery Institute (*Explore Evolution*, 2007), although written by leaders of the intelligent design movement, does not mention ID anywhere, not even in the index, presumably to distance the authors from the failure of intelligent design in the Dover, Pennsylvania, trial. Omitting ID likely would also help the book avoid constitutional scrutiny. Thus, the current strategy is to back away from trying to insert ID and instead work toward criticizing and weakening the teaching of biological evolution, remaining silent about the alternatives they hope to insert into public science classrooms.

Somewhat ironically, creationism continues to evolve.

# 10

# Is There Evidence Supporting Intelligent Design?

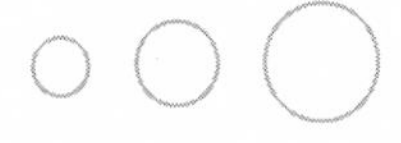

*Easily the biggest challenge facing the ID community is to develop a full-fledged theory of biological design. We don't have such a theory right now, and that's a real problem. Without a theory, it's very hard to know where to direct your research focus. Right now, we've got a bag of powerful intuitions, and a handful of notions such as "irreducible complexity" and "specified complexity"— but, as yet, no general theory of biological design.*

PAUL NELSON

*When ideas fail, words come in very handy.*

JOHANN WOLFGANG VON GOETHE

It's important to take a closer look at the arguments that people have used to try to support the teaching of ID as if it were a science. These arguments are usually not presented as evidence *for* something as much as they are complaints *against* problems they perceive in evolutionary theory. One line of reasoning and evidence (evolutionary theory) is attacked, but no real alternative is supplied. Another tactic (which is not unique to creationism or ID) is to use an "either/or" approach: if doubt can be cast on idea number one, then by default idea number two must be correct, as though these were the only two possible alternatives. It's a setup, a "contrived dualism,"[1] where any evidence against evolution must be evidence in favor of ID. That's silly, of course; it's like saying that evidence against a flat earth means it must be shaped like a donut. In addition, the either/or approach makes it sound as though there is a single alternative to biological evolution, when in reality creationism has a wide variety of forms, including young-earth creationism, progressive creationism, intelligent design, and others. Even within the intelligent design movement there is a great range of positions, from young-earth believers (e.g., Paul Nelson) to those who believe in a 4.5 billion-year-old earth and an evolutionary origin of all species, including humans, by way of natural selection (e.g., Michael Behe), prompting us to ask what, exactly, is being offered as the alternative to evolutionary biology.

The perceived problems in evolutionary theory pointed out by ID supporters include "gaps" in the fossil record (also a favorite argument of creationists); asserting that life (or parts of life) are simply too complex to have arisen by evolutionary processes; pointing out that some species or groups of species have not changed much over time (which is true); pointing out disagreements among evolutionary biologists; or expressing the fear that evolution somehow removes God from the picture altogether—harking back to our prism and rainbow example. Let's take a closer look at some of these arguments.

As noted, some ID advocates point out disagreements among evolutionary biologists as evidence for questioning evolution altogether. There certainly are disagreements among evolutionary biologists, sometimes heated ones. But these disagreements have to do with ongoing investigations into the mechanisms and patterns of evolution, not with evolutionary theory itself. Disagreements and ongoing discussions are a sign of any healthy and vibrant field of science; they do not imply that there is a "theory in crisis" as is sometimes charged. Indeed, the disagreements and continued refining of our knowledge is what makes a science a science—we don't know everything yet, and we are continuing to explore and to discover.

Then there is the fear that evolution somehow removes God from the picture—a fear that usually is not voiced in court cases or with school boards so that the arguments will not appear religiously motivated. It is mostly the "randomness" of mutations that seems to concern creationists here, since this would seem to imply pure chance over the hand of a designer. But randomness in the context of genetic mutations simply means that they are unpredictable, as is true of so much about life. We simply do not know which variations will occur, when they will occur, or in which organisms. Additionally, "random" in this context means that mutations occur regardless of whether they are correlated with fitness (i.e., regardless of whether they are helpful or harmful to the organism).

So far, I've talked mostly about the various attacks on evolutionary theory, and I have not mentioned any "positive" lines of evidence in support of ID theory. Do any exist? Yes, there have been ideas put forth in support of ID. But these ideas are not "science" in that they are not testable, they have been severely criticized by scientists as well as by theologians, and they have not led to any productive lines of research. The best known of the ID arguments is the concept of "irreducible complexity," the idea (expressed mostly by biochemist Michael Behe) that many living things, or at least parts of them, are so amazingly complex that they could not have arisen from natural selection and other evolutionary processes over time. The slow accumulation of minor changes over time would not, in their opinion, explain these complex structures and organisms, and any "intermediate" forms would not have worked. The individual components of these systems do not work on their own, and so they could not have been subject to natural selection; only the entire assembled "machine" would be. It's a very appealing argument, since living systems are so amazingly complex. The appeal lies partly in the "what good is half a wing" argument: How could a wing have arisen by small steps through natural selection when the intermediate steps (half a wing) would not have worked? Behe's contribution has been mostly to make the same argument on a cellular and molecular level.

The irreducible complexity argument is not a bad one if you make the comparison (which they do) with man-made structures such as watches, bridges, buildings, and other complex things that we ourselves designed. But it falls apart when looking at biological structures. It's relatively easy to see (and to test) methods by which the mechanisms of plants and animals have arisen over time, though admittedly this is still an active field of science and there is much we do not yet know. Behe has written about the intricacy and complexity of the whip-like flagellum that propels certain bacteria, and this structure has become a favorite piece of "evidence" in favor of irreducible complexity. It was even featured in the Dover, Pennsylvania, court case, discussed later in this chapter. But the bacterial flagellum, an admittedly complex and awe-inspiring structure, has been shown to have been cobbled together over time from pre-existing and functional cellular components.[2]

And, in general, there *is* great value in having half a structure: a poorly developed, stubby wing might allow you to at least sail or glide; feathers insulated the body before aiding in flight; a light sensitive organ that is not as complex as the mammalian eye could at least show you the light. The "what good is half a structure" argument supposes that natural selection has some sort of goal in mind, and that anything short of that ultimate goal would not work. Instead, natural selection operates on features that already exist, and this makes the irreducible complexity argument easy to refute. And the "contrived dualism" problem is here as well: simply casting doubt on the efficacy of natural selection does not by itself point to a designer.

A further problem that I have with the irreducible complexity argument is that it seems to close the door on asking any further questions. If something is too complex, well, that's the end of it. In a way, it's like saying that if we cannot see how it could have happened biologically, then it must have been the hand of God, and surely God could not have done this through natural selection, since we cannot imagine how that might have happened. And so once again we're back to the "God of the gaps" thinking, where God is confined to the shadows of our ignorance.

ID proponents will also sometimes refer to the argument of "specified complexity," an idea advocated mostly by philosopher and theologian William Dembski. Something is said to have "specified complexity" if it appears to be both complex and specified, which according to laws of probability would point to the hand of a designer rather than to natural selection. But that concept has been severely criticized, as it is at best vaguely defined, even by its defenders, and it has not led to any new findings or further work. It cannot be tested, it is not science, and so it has no place in a science classroom. It is easy to point to things that are extremely complex—nature is full of them—but simply saying that something is complex does not tell us anything.

Additionally, the ID advocates refuse to say *when* all of this designing must have taken place. Consider, for example, the major Hawaiian Islands, which did not exist until some 5 million years ago (the smaller islands to the northwest are much older). These islands are full of plants and animals found nowhere else. Has there been a separate "design event" to account for their impressive diversity of unique species? And does this happen whenever another island is formed, whenever one species becomes extinct, or whenever a new species appears? It's been estimated that more than 99.99% of all species that ever lived on Earth are now extinct, with these extinctions occurring continually throughout history, including some mass extinction events from time to time. What does this say about God as designer? That God is indecisive or fickle? That God is incapable of producing a good design, one that lasts more than a few million years? The ID proponents are frustratingly (and it seems to me intentionally) vague on these points.

The fact that "chance" bothers them so much (and seems to mean "godlessness" to them) intrigues me. Does chance mean the absence of God? Does not God operate through chance in the Bible in the casting of lots (e.g., Leviticus 16:8; Joshua 18:6, 18:8; Jonah 1:7), even with regard to selecting the apostles (Acts 1:26)? Chance is everywhere in nature, and there are many who argue for chance being an integral part of God's divine plan.[3] If chance means the absence of God, then surely God must be absent from any card game or state lottery where chance is involved, not to mention predicting the weather. Are we being told where God is? The ID advocates, like the creationists, appear to be comfortable telling the rest of us where God is and is not. Possibly without realizing it, they are essentially placing limits on what they think God can, and cannot, do. It is essentially an unwillingness to admit that God is capable of all things, not just those few we can easily understand.

This line of reasoning is a sad commentary on the inquisitive nature of mankind in general. In the face of any problem that appears to be too challenging, are we to throw up our hands, stop asking questions, and invoke God? Michigan State University philosopher and science historian Robert Pennock summarized this problem as follows: "Once such supernatural explanations are permitted they could be used in chemistry and physics as easily as creationists have used them in biology, geology and linguistics. Indeed, all empirical investigation could cease, for scientists would have a ready-made answer for everything."[4] Where would we be if, throughout history, we had avoided or abandoned all difficult questions, if we had responded to all roadblocks to knowledge by simply giving up? I am not equating "giving up" with seeking God; I am saying that we are expected by God to use our gifts of reason. Theologians and scientists have recognized this for centuries.[5]

When the argument that ID is some form of science fails, as it has done so often, the ID advocates sometimes turn to the argument that evolution, a naturalistic process, is somehow atheistic. Therefore, they imply, teachers of evolution are just forcing state-sponsored "atheistic naturalism" onto unsuspecting school children. It's a weak way of saying that if ID is religion and cannot be taught because of the First Amendment, then evolution also violates the First Amendment because it violates the U.S. constitutional requirement of religious neutrality. In order to make this claim, they try to portray evolutionary science as an anti-religious "worldview," and they then give it names such as "Darwinian orthodoxy" or "dogmatic Darwinism" to argue that it's an atheistic worldview rather than a science.[6]

This "atheistic worldview" charge is very curious. First, it's an assumption that the ID advocates know the religious views of everyone else, including the teachers who teach evolution. Someone who is an unabashed defender of the science of evolution becomes not only a bad guy in their eyes but also, what is worse, a devotee of "dogmatic Darwinism," described by them as a dangerous worldview that deprives life of its meaning and purpose, undermines the morals of young people, threatens to take over our social structure, and tosses out God. Second, by arguing that evolution is atheistic, but then turning around and saying it must be "balanced" by offering ID, they betray the fact that ID proponents themselves really know that ID is a religious belief. And of course, if evolution were truly atheistic, that would mean that all of the other sciences must be atheistic as well, since all of them—chemistry, geology, molecular biology, astronomy, physics, geography, and cell biology, to name a few—clearly support, and are unified by, evolutionary theory.

Most ID creationists do not even allow the religious viewpoint of "theistic evolution," the belief that God employed evolution as the mechanism by which to shape life on Earth (as raindrops and light may be used to form rainbows), a view held by a large number of Christian denominations.[7] The point of their arguments seems to be that all over the world there is some sort of evil plot[8] among scientists and science educators, but especially among evolutionary biologists, that is being hatched against misunderstood Christians who simply want fair treatment. Clearly the attempt to paint evolutionary science as an atheistic worldview is the weakest of their arguments. It is an almost childlike attempt to put ID and evolution on equal footing.

But ID and evolution are not on equal footing. Intelligent design, just like "creation science," is religion masquerading as science,[9] masquerading much better than in the past, perhaps, but masquerading all the same. Advocates of ID will say that their work only points to a designer, and that the designer need not necessarily be the Judaeo-Christian God, since to say that would be unscientific. But there are few candidates for the job of designer, and everyone knows that they are referring to the God of Christianity. The major proponents of ID—Phillip Johnson, Michael Behe, William Dembski, and others—are Christians who readily admit that they picture, as the designer, the one God of Christianity. There is certainly nothing wrong with that, but when that fact is exposed it becomes quite clear that the promotion of ID is a religious movement with a religious and political agenda; it has nothing to do with science or science education.[10] Evolution, in stark contrast, is a valid and vibrant field of scientific inquiry, studied, taught, and understood by a wide variety of people of all faiths, including Christianity, around the world. Evolution is not *better* than ID, and ID is not better than evolution; one is simply science, the other an odd form of religion.

The most recent attempt to insert the teaching of intelligent design as an alternative to evolution in an American public high school science curriculum was soundly defeated in federal court. In December 2005, after a six-week trial that garnered national and international attention, U.S. District Judge John E. Jones III delivered a clearly written and well-reasoned decision against the attempt to teach "intelligent design" (which he recognized as a relabeling of creationism) in the high school classroom in Dover, Pennsylvania. Judge Jones employed some of the most forceful language used to date in any such case, calling the attempt by the Dover school board to insert creationism/intelligent design into public schools "breathtaking inanity" and a "sham," in addition to being a clear violation of the U.S.

Constitution.[11] Teachers, scientists, and religious leaders all over the world applauded this decision.[12]

This rational, logical, and strongly written decision by Judge Jones, a conservative, Christian, Republican judge who was personally appointed by President George W. Bush, should have been the end of the issue. But, of course, it wasn't. Early in 2006, a similar controversy erupted in the small mountain community of Lebec, California.[13] And anti-evolution legislation was introduced early in 2006 in Utah, Oklahoma, Mississippi, and Alabama. The National Center for Science Education, an organization that tracks such activities to keep the public informed of attempts to weaken the teaching of science, reports that in early 2008 nine county school boards in north Florida adopted resolutions calling for "equal time" teaching of "alternative theories" in science (to open the door for creationism), and the Institute for Creation Research, a young-earth creationist organization, requested permission to issue a master's degree in science education in the state of Texas. Anti-evolution bills were also considered in Alabama, Michigan, Missouri, and South Carolina.[14] In Louisiana, S.B. 733, the "LA Science Education Act," was signed into law by the governor in 2008, despite being recognized by teachers and scientists as "stealth creationism." The act was supported by the Discovery Institute (the Seattle headquarters of the ID movement) and by LA Family Forum (a pro-creationism Louisiana affiliate of James Dobson's Focus on the Family ministry).

Rather than openly promoting ID, the recent tactics involve more general antievolution activities using language such as "strengths and weaknesses" and "academic fairness" in order to open the door to get ID materials into classrooms. Somewhat predictably, the new bill does not argue for the insertion of creationism or intelligent design, but instead argues for "critical analysis" that would allow teachers to deviate from approved science curricula in three areas that Louisiana creationists and politicians apparently find troublesome: cloning, evolution, and global warming. No other scientific topics are mentioned.[15] Why would only these three topics, among the many fascinating fields of science, be perceived as being in need of some kind of special attention and critical analysis?

# 11

# Human Arrogance

*The arrogance of men will be brought low and the pride of men humbled.*

PSALM 2:17

*I distrust those people who know so well what God wants them to do because I notice it always coincides with their own desires.*

SUSAN B. ANTHONY

Although I am a scientist, I do not know anyone personally who has seen a molecule, or an atom, or an electron. Indeed, although we can observe indirect images of atoms using the most sophisticated tools available (scanning tunneling electron microscopes), the best we can do is to infer their presence. The electron especially both intrigues and worries me. If, as has been described by some physicists, the electron is simultaneously a wave and a particle, has almost no mass or weight, and cannot be seen, how do I really know that it's there? This seems to me like science at its most questionable, should someone want to question it.

And yet I have never heard of a serious public debate over whether electrons exist. I have not heard of a public debate over the depths of the oceans or how we know those depths, or the shape of the earth (at least not in the last hundred years or so), or how a virus operates. I have not heard of any public debate over how we calculate the distance to the nearest star, or how we measure the heat of suns in galaxies not our own, things that to me seem truly astounding and worthy of raising the question "how do you guys know that?" But evolution seems to be fair game, and it never ceases to amaze me that people with absolutely no background in biological or geological sciences and very little understanding of science in general feel quite comfortable standing up to say that evolution simply cannot be so.

Why is this? It's primarily because nobody really feels threatened (or feels that their faith may be threatened) by electrons, or oceanography, or astrophysics. But they do feel threatened by evolution, because it somehow lessens their sense of importance to be linked to other forms of life on Earth. Polls have shown that many people *understand* evolution but simply balk at the thought that they might be *products* of evolution.[1] I have always suspected that some of this confusion stems from the verse in Genesis (Genesis 1:27) where "God created humankind in his image, in the image of God he created them," a passage that some readers of the Bible take literally. The footnote to this verse in my Bible indicates that the "image" referred to here was intended to be a reflection of God's grace and mercy rather than any physical likeness. But not everyone interprets it that way.

The tendency to object to any suggestion that we might be somehow related to all other life on Earth is a result of arrogance. It comes from the feeling that we are so special that the entire universe revolves around us (quite literally, in days of old). Creationists (including ID advocates) believe—or rather, they think that as Christians they are *supposed* to believe—that any evidence linking the diverse forms of life on Earth must be wrong, since to them it contradicts their interpretation of the Bible and makes them feel less special. We *are* special, and there *are* differences between humans and other species, such as our ability to create art, compose music, communicate via complex languages, and, unfortunately, display arrogance. What is sad about the anti-evolution, anti-intellectualism arguments is not just that they are wrong, but that the arguments are created in the mistaken belief that it is "the Christian thing to do." One cannot help being reminded of the geocentrists, flat earthers, and snake dancers here, all of whom were similarly convinced.

The Galileo incident, where the great Italian physicist and astronomer Galileo Galilei was found guilty of heresy based on his scientific observations, is a nearly perfect example of human arrogance and is one of the better-known cases of science and religion in conflict. I say "nearly" perfect because the tale, as usually told, is a bit of an exaggeration. Galileo's eight-and-a-half years of imprisonment were, in fact, "house arrest," and he was eventually allowed to continue his scientific writing unimpeded by the Roman Catholic Church. And in fact, Galileo enjoyed the support of various Church leaders, including the pope, throughout most of his career. The Church was not as ignorant, and Galileo not as innocent, as most traditional versions of the story relate. Additionally, at least some historians suspect that the situation might have been avoided had Galileo himself been somewhat less arrogant in his dealings with the Church, and there were actually astronomers and Church officials on both sides of the issue. But the basics of the story hold true.

Galileo was a deeply religious man who was also a scientist with keen powers of observation. Had he simply pointed out that the moons of other planets moved (supporting the Copernican view of the universe), chances are nobody would have cared too much. He would have been just another scientist making interesting but arcane observations. But when Galileo insisted that the earth itself was moving—meaning that we might not be at the very center of God's world—many Church leaders (and also, in fairness, some astronomers of the day) went ballistic. Because of Galileo's writings on the subject, a committee of advisors to the Roman Inquisition declared in 1616 that Galileo's belief that the earth circled the sun was absurd and heretical.[2] Galileo was eventually summoned to appear before the Inquisition, where he was interrogated, then arrested, forced to confess that his views were overstated, and sentenced to confinement.[3] Most of his accusers knew little about astronomy. They knew only that it bothered them to be thought of as being anywhere but at the center of God's world, a view that was, according to their interpretation, clearly supported by Holy Scripture in abundance. The parallels to creationism and its successor, the ID movement, are all too clear.

It's not difficult to find similar cases of arrogance getting in the way of our progress. As noted earlier, like creationists, flat-earth believers were absolutely convinced that they alone knew the truth; their understanding of the Bible was the only correct position possible: the earth simply had to be flat. Their arrogance kept them from even listening to, let alone understanding, the truth about the world around them. In the field of biology, the curious affair of Trofim Lysenko also comes to mind. Lysenko, a scientist who denied Mendelian genetics in favor of crackpot schemes and wild "successes" that no one else could duplicate, irreparably damaged science and science education in the former Soviet Union for nearly half a century. Appointed head of the Soviet Academy of Agricultural Sciences because his views were seen as politically correct at the time, Lysenko was responsible for the expulsion, imprisonment, and death of many scientists who studied genetics and for basically outlawing any Mendelian genetic research throughout the entire Soviet Union.[4] The arrogance of the USSR political system in their inability to admit their misplaced trust, and of Lysenko himself, excluded any possibility of admitting error.

Scientists have often been accused of being arrogant, or elitist, or both. And some of them are, take my word for it.[5] Those that would use science to argue against the existence of any God, something that science cannot, by definition, even begin to investigate, certainly seem to me to fall into that category. But arrogance is not exclusively theirs.

And when it comes to debates over evolution and creationism/ID, I've seen very little to match that of the ID proponents. It's difficult for me to imagine climbing onto a stage to debate a topic about which I know very little and have done no in-depth research or study. Yet repeatedly, they step up to a podium or create detailed web sites to declare that evolution simply cannot be. It's important to understand the message underlying their actions: They have decided not only what they want to believe, but also what you should believe. They have decided—although they will not say so in courts of law or in public school board meetings—that their interpretation of the Bible is the correct one, and that it should also be your interpretation. Not content to stop there, they want to mandate that their view is taught to children in a public school, supported by our tax dollars, and disguised as a form of science. They have decided that they speak for all Christians. Perhaps without even realizing it, they have decided that they, and they alone, understand the mind of God.

# 12

# In the Beginning

*In the beginning, when God created the heavens and the earth, the earth was a formless void and darkness covered the face of the deep, while a wind from God swept over the face of the waters.*

GENESIS 1:1–2

*Come together, right now, over me.*

LENNON & McCARTNEY, 1969

By far the most frequently mentioned passages in the Bible in discussions and debates about creationism, science, and evolution are the early verses of Genesis. The opening words quoted above are as familiar as they are awe-inspiring and comforting to millions of people worldwide. The King James Version has a slightly different translation of the last line: ". . . And the spirit of God moved upon the face of the waters." I have always thought that these opening lines were hauntingly beautiful. In fact, I have often wondered if my fascination with water harks back to hearing these words read to me early in life. As a child, I developed an image of God as a great wind moving over the face of the waters in the same way that it blows through us, and that image remains with me today. This might even explain why I chose to study life in the sea, and why I cannot ever seem to get enough of fly fishing. Right or wrong, in my mind water and God are somehow inextricably linked. I wonder if others feel that, and if that's partly what baptism is about.

Although Christians regard the Bible as being divinely inspired, nearly all serious students of the Bible are aware that Genesis was written by different people at different times. For instance, most people are aware that there are actually two fairly different creation stories in Genesis 1 and 2; they differ markedly in their style, order, facts, and choice of words. In Genesis 1, for example, God is always referred to as Elohim, whereas the writer of Genesis 2 never uses that word, instead referring to God always as Yahweh (YHWH). Much has been written about these differences,[1] and whether these two stories contradict or complement each other, and the problems that arise when readers attempt to reconcile the differences or view Genesis as a scientific text. That's not my purpose here. Instead, I want to address an issue that I don't think has received much attention to date: the fact that Genesis, and indeed the entire Bible, actually points toward, rather than away from, an evolutionary understanding of life on Earth.

What do I mean by this? If someone were to ask you if the overriding message throughout the Bible was one of unity or one of division, what would you say? In both the Old and New Testament, is the dominant theme one of separation or one of coming together? Genesis, in describing the creating of life by God, does not say anything about the *relatedness* of those forms of life, only that they were all creations of a loving God. There is nothing in Genesis that argues against the concept of a single, unifying tree of all life.[2] It is true that there is a lot of "separation language" in Genesis—separating light from darkness, morning from evening, water from sky, land from sea, even creatures separated "according to their kind" (and interestingly, this last creating is done not directly but through natural processes; God orders the *earth* to bring forth living creatures [Genesis 1:24]). But the overarching story of Genesis is about the relationship between God and mankind; it is a story of the overall "oneness" of creation, of all life having a common origin, explained in the only terms that would have made sense to the ancients. We come from One, and so we have a great yearning within us for "oneness," for meaningful relationships and connectedness in an often-fragmented world.[3]

This concept of the oneness of all life is even more clear in the New Testament. In Paul's letters to the churches of Corinth and Ephesus he refers to the sacred connectedness of all humans when he uses the analogy of a single body consisting of many parts, and where by one Spirit we are all baptized into one body (e.g., 1 Corinthians 12:13–27 and Ephesians 4:4–6: "There is one body and one Spirit, just as you were called to the one hope of your calling, one Lord, one faith, one baptism, one God and father of all, who is above all and through all and in all"). Paul was using the human body as a metaphor for the body of Christ. An evolutionary view of the world merely extends Paul's analogy to all the rest of creation. In the last-written of the four gospels, Christ's concern about us coming together is also clear: "I ask not only on behalf of these, but also on behalf of those who will believe in me through their word, that they may all be one"(John 17:20).

The "oneness" of the Bible, the weaving together of so many stories, parables, poems, and songs into a single coherent message, is one of the great triumphs of literature, and it is for Christians a cause for amazement and celebration. In the words of Australian priest Denis Edwards, writing as a follow-up to a conference on evolution and divine action, "The diversity of life on Earth, interconnected and interdependent in the one biosphere of our planet, is a sacrament of divine Wisdom."[4] This sense of oneness is also what is so tragically lost in fundamentalist approaches to faith (any faith). Today's fundamentalists, and some of today's evangelicals, by asserting that what is holy is contained only within the inflexible walls and narrow confines of a small and pre-defined space, are promoting fragmentation and division. That view is in stark contradiction to the message of unity and togetherness, and of the removal of barriers between us, that are so clearly expressed in the New Testament. "For he is our peace; in his flesh he has made both groups into one and has broken down the dividing wall, that is, the hostility between us" (Ephesians 2:14).

Science does not contradict this sense of "oneness." Quite the opposite, in fact. Our current understanding of the beginning of the universe points to a single time and place (commonly referred to as the Big Bang) from which all matter in the universe, having been previously amazingly condensed, exploded outward and became the expanding universe we know today. Today's theoretical physicists are searching for universal laws that would unify many of our most basic physical laws and particles of nature (sometimes referred to in popular accounts as "Superstring Theory" or the "Theory of Everything"). Evolutionary biologists have amassed overwhelming evidence that life, too, had a single origin in the distant past, a finding that explains the relatedness of all life on this planet. This sense of connectedness is also what environmentalism, our call to care for the planet, is all about; we understand now that each of the components is part of a larger whole: "If one part suffers, all suffer together with it; if one part is honored, all rejoice together with it" (1 Corinthians 12:26). I am not saying that these disparate fields of scientific study point to the same moment in time, nor am I saying that these findings support a literal reading of Genesis; I am saying only that nature seems always to remind us about the connectivity of life on Earth.

The overarching message throughout the Bible is one of unity rather than separation, of coming together rather than of tearing apart, of relationship of one to another. This unity, this "oneness," extends to all of creation. In my view, the realization that all life on this planet shares a common origin and is related—that is, the realization that biological evolution has occurred and continues to occur—could not be more in accord with this message.

# 13

# The Unnecessary Choice

*If the world is not God's, the most eloquent or belligerent arguments will not make it so. If it is God's world, and this is the first declaration of our creed, then faith has no fear of anything the world itself reveals to the searching eye of science.*

THE RT. REV. BENNETT J. SIMS,
EPISCOPAL BISHOP OF ATLANTA

The Galileo incident and the history of belief in a flat earth[1] are fairly well known. But these are both exceptions to the rule rather than typical examples of the interaction of science and faith. Throughout history, faith and science have actually coexisted quite nicely. Indeed, it has only been in relatively recent years that the two have been clearly separated. As recently as the early nineteenth century, what we today call "science" was really more of a general "search for truth" that employed religious as well as natural explanations and observations. And although some incidents are memorable for being pretty nasty, there have always been respected theologians welcoming the advances of science, and there have always been competent scientists with a deep and abiding faith. Yet for some reason, the issue of science and faith continues to be seen by many people as a choice that has to be made.

A well-liked and respected youth leader recently addressed a mission group working to improve a poverty-stricken region in the Appalachian Mountains. The mission group contained a large number of high school members from my church. By all accounts, he was a deeply committed Christian, a wonderful speaker, and a dear and genuine person. The kids loved him. Among his many heartfelt and meaningful messages was the following sentiment: "I believe that the Grand Canyon was created by God, and not by some river."

That statement is representative of Christians who feel that they have to make a choice between science (geology, in this case, which indicates that the Colorado River played a major role in the formation of the Grand Canyon) and God. I have heard similar statements—implying choices that have to be made—about a variety of natural phenomena, but evolution is by far the most common target. One evangelical Christian web site (called Stand to Reason) has the following rather bleak statement about evolution (called Darwinism here following the current fad among anti-evolutionists) and ends by posing a question designed to force the reader to make some sort of choice between science (in this case evolution) and God:

> If Darwinism is true, then there is no purpose or meaning to life, there is no morality, there's no qualitative difference between humans and animals, there's no life after death, and there's no purpose to human history. Now, are you trying to tell me that it doesn't really matter if people believe we evolved or not?[2]

I am saddened to think of the number of visitors to that site (and to many other web sites like it, some far more antagonistic) who will read it and come away feeling that a choice has to be made, with

God on one side and science on the other. I'm sure the intent of the statement is not to drive students away from Christianity. But when students encounter serious evidence for various sciences as they attend colleges and universities, and then feel that they must make a choice based on statements like the one above, it's clearly a lose-lose scenario—they will either drop their faith or drop science.

As is the case with other examples I have presented, this is, for some people, an honest mistake. That is, there are a large number of well-meaning people who are simply trying to do what they think is right. For these people, when a choice is presented in such stark terms, with God posited on one side only, it's no surprise that they "choose God." For other people, there is some intentional dishonesty involved; that is, there are some writers and speakers who continue to present the issue as a choice between good and evil, black and white, faith and science. They know better. And they prey upon the good intentions of the large number of people who are honestly searching for answers.

A recent spoof on the web (printed in the online version of *The Onion*[3]) refers to the theory of "intelligent falling" as an evangelical replacement for gravity, which is described in that article (a satire of intelligent design) as a "theory in crisis" because we do not yet understand all of the details of gravity, especially as it relates to quantum mechanics. Beyond the obvious humor they are intending, the article makes the point that for people who see things in this black and white approach to Christianity, a choice needs to be made: either God or gravity, but not both.

The implied "choice" between science and faith is as unfortunate as it is unnecessary. Science and faith are not competing hypotheses of how the world works. They are not meant to be, and to argue one over the other is to miss the point of both. As British literary critic Terry Eagleton put it, "the difference between science and theology is, as I understand it, one over whether you see the world as a gift or not; and you cannot resolve this just by inspecting the thing, any more than you can deduce from examining a porcelain vase that it is a wedding present."[4]

Newton's prism, you will recall, did not create the colors contained in sunlight. It simply revealed them. Science does not invent nature; it simply reveals nature. If this is God's world, then science can only reveal God's world. The formation of rainbows by the prismatic interaction of light and water—revealed by the science of optical physics—does not remove God from this beautiful spectacle. The formation of canyons by rivers—revealed by the science of geology—does not remove God from the Grand Canyon. The fact that the earth revolves around the sun rather than vice versa—revealed by the science of astronomy—does not remove God from our universe. The relatedness among all forms of life on Earth, humans included, and the diversification of life over millions of years—revealed by the science of evolution—does not remove God from the history of life, or from our lives.

# 14

# What Are We to Believe?

*For every thing there is a season, and a time for every matter under heaven.*

ECCLESIASTES 3:1

*If we are children of God, we have a tremendous treasure in nature and will realize that it is holy and sacred. We will see God reaching out to us in every wind that blows, every sunrise and sunset, every cloud in the sky, every flower that blooms, and every leaf that fades.*

OSWALD CHAMBERS

So what, as Christians, are we to do? What are we to believe?

I would recommend several things. First, if you are a Christian or are even mildly interested in learning more about the Christian faith, I think it's worth the time it takes to really study the Bible. Don't take my word, or anyone else's, for what's in there. Check it out for yourself. The Bible is an incredibly complex book, written at different times by numerous people in diverse settings, and it is full of priceless lessons. It is not a textbook; it is a collection of enduring messages passed down from generation to generation. It is, above all else, an invitation.

A wise reader of the Bible does not get caught up in the details and lose the important message. I find the Bible's history, the context in which the various books were written, and what little we know about the authors themselves to be fascinating. There are reasons behind, as well as messages within, these exquisite stories. Even apart from any religious consideration, there is no arguing that, as concerns Western civilization, this is the most influential book or collection of books ever written. It is deserving of study for that reason alone. It is worth reading individually as well as studying in group settings or in classes with people who know it well, people who have devoted their lives to its analysis and interpretation. Listen to the learned, but also use your own mind and heart.

Second, I don't think that there is any need for you to question your faith just because you perceive a conflict between the Bible and things revealed by science. Questions and doubts are good things, and learning is a good thing, and we're expected to struggle with difficult issues and to use our minds. Can the simple teachings of Jesus be followed with no concern whatsoever that the findings of modern science will lessen your personal religious beliefs? I think so. The light of science and the light of faith are meant to be illuminating, not blinding. And like other sources of light, the two will complement, and not negate, one another. No finding of science need ever lessen your capacity or your need for worship. You can devote your life to the things of lasting value so clearly outlined throughout the Bible: helping others, dispensing kindness, showing tolerance, giving thanks, promoting peace, and worshipping. But you'll want to do so intelligently, ". . . with all your mind," celebrating God's gift to you of clear reasoning, and knowing he is capable of all things, not just the few things we can easily understand.

Third, if you are really interested in understanding the biological sciences, including evolutionary theory and all that it entails, then you should study it. Learn the principles that so clearly demarcate science from non-science to understand why evolution is a major branch of modern science. There are excellent books and journals devoted to this most vital of the life sciences. You will find the literature vast, daunting in both volume and detail, and very exciting. If possible, contact a professor in the biological sciences who studies and teaches evolution and accompany him or her on field trips or in the lab. There is no substitute for actually knowing something about a topic.

Finally, don't fear knowledge, or the way that we generate most knowledge. Science, which I see as a gift from God, is the most direct way we have of learning about the natural world. And although some people (including some scientists) have come off as being hostile to religion, science itself is not remotely anti-Christian or anti-religious. Nothing we can discover about this world will come as a surprise to God, and we have nothing to fear from learning about the world around us.

# Epilogue

*There are only two ways to live your life. One is as though nothing is a miracle. The other is as though everything is a miracle.*

ALBERT EINSTEIN

My time spent as a youth worker in the Presbyterian Church has been one of the most enjoyable and rewarding experiences of my life. Students ask some very challenging questions. One that has come up fairly regularly is: Where is the evidence of God in the world today? We read about his works and his miracles in the Old and New Testaments, and they sound amazing and wonderful, but these miracles—from the burning bush and plagues of Egypt to Jesus's turning water into wine, healing the sick, raising the dead—all happened 2,000 or more years ago. What proof do we have that God is alive and with us now?

Perhaps one way to look at evolution is as the answer to that vexing question. What more tangible, palpable evidence could a Christian ask for than to know that the earth's species—ourselves included—are still changing, still being modified, still evolving? To know, in other words, that God's work is still being done? And that the evidence is right here in front of us, staring us in the face, in the form of clear, unambiguous biological, physiological, geological, physical, chemical, astronomical, and paleontological data, easily available to us through the lens of science? The "random chance" part of life that is so frightening to the creationists and ID advocates should be seen in a far different light: Despite the incredible odds against it happening, despite the fragility of life, despite the vagaries of biological evolution, occurring as it did over millions of years, with all the attendant mutation, recombination, luck (chance), selection, and more—we *are* here. And surely, in anyone's book, that should fall under the heading of miraculous.[1]

Rather than looking at evolution as a conspiracy or invention of evil scientists, which it clearly is not, or as a threat to your faith, which it also is not, look at it as you look at everything else that is part of this wonderful, incredible world around you, a world waiting to be discovered, a world that we are instructed to appreciate and understand repeatedly in Holy Scripture: look at it as part of God's world.

# Acknowledgments

Many people, not all of whom have agreed with, or approved of, my style, presentation, or conclusions, have read over parts of this book at various stages, and I thank them sincerely for it. They include Andrew Lustig (Davidson College), Michael Ruse (Florida State University), Wes Elsberry (National Center for Science Education), Jarvis Streeter (California Lutheran University), Ken Miller (Brown University), Kirk Fitzhugh (Natural History Museum of Los Angeles County), Barbara Forrest (Southeastern Louisiana University), Linda Silver (Great Lakes Science Center), Kevin Padian (University of California Berkeley), David Steinmetz (Duke University Divinity School), Dallas Willard (University of Southern California), William McComas (while at the University of Southern California), John F. Haught (Georgetown University), Keith B. Miller (Kansas State University), senior editor Vincent Burke and senior production editor Mary Lou Kenney (Johns Hopkins University Press), and the following people who are now or were formerly associated with Westminster Presbyterian Church in Westlake Village, California: John Burnett, Rob Douglas, Nathan Reeder, Rob Seitz, Dick Thompson, Suzanne Thompson, Beau Wammack, and Andrew Cahill. I thank the Rev. Paul W. Egertson (California Lutheran University) for pointing me toward more accurate estimates of some of the membership numbers that appear in the appendix.

I am especially indebted to four outstanding youth workers, Rob Seitz, Dave Carpenter, Rob Douglas, and Nathan Reeder, for providing me with the opportunity to interact with middle school and high school students in a welcoming, learning, and supportive church setting. I have also benefited greatly from conversations with students, friends, and relatives during recent years, among them Matt Toyama,

J. Trent Collier, Roger Ellis, Scott Adam Daehlin, Don Partenfelder, and the always-inspiring students of the Westminster Presbyterian Church SYF program.

Most of all, I thank my wife, Sue, for her constant love and support and my children, Alex and Paul, for their inspiration. All three of them also read parts of the book as it was being readied for publication and offered helpful suggestions.

# Appendix

## MAJOR CHRISTIAN DENOMINATIONS AND THEIR STANCE ON SCIENCE, EVOLUTION, AND CREATIONISM / INTELLIGENT DESIGN

My assertion in the introduction and chapter 1 that most Christians accept biological evolution as being compatible with their faith is easily supported by referring to the Catholic Church. Catholics (including Roman Catholics and Eastern Orthodox) constitute some 1.2 billion of the estimated 2.2 billion people worldwide who count themselves as Christians. That's more than half of all Christians and roughly one-sixth of the entire world population. Apart from the Galileo episode, the Catholic Church has had a long history of accepting the findings of science, including evolutionary biology, as revelations of how God's world works.

However, because the creationism and ID movements are very much a Protestant (and mostly U.S.) phenomenon, we should look beyond the Catholic Church. In the United States, there are slightly more Protestants than Catholics[1] (approximately 52% to 44%, according to one estimate).[2] Protestant groups are more divided on the issue (see below and table 1 [in chapter 1]). To create this appendix, I searched for statements about science and faith (or evolution and creationism) from some of the larger Christian groups and movements (mostly Protestant) in the United States, and in some cases where I could find no public statement, I sent queries to their national headquarters. These groups are listed below. Where feasible, I have included the same basic information for each group: the official name of the organization or denomination; the organization's approximate size (estimated number of members) in the United States;[3] its position on the compatibility of its members' faith with evolutionary biology, usually in the form of publicly available statements; the dates of those statements; key phrases from the statements; and (in some cases) the name of the person responsible for making the statements. For many religious organizations, no such statements could be found.[4] Non-trinitarian groups (e.g., Jehovah's Witnesses, Church of Jesus Christ of Latter-day Saints, Unitarian Universalists) are not listed here, though the latter two are supportive of evolution being compatible with faith.

## CATHOLIC AND EASTERN ORTHODOX CHURCHES

### Roman Catholic Church (66.6 million [62.0[T]–71.2[A]])*

Recent Statements

- 1996. Pope John Paul II, speaking at the annual meeting of the Pontifical Academy of Sciences, the Catholic Church's "science senate."

> "Today, almost half a century after the publication of the Encyclical, fresh knowledge has led to the recognition that evolution is more than a hypothesis. It is indeed remarkable that this theory has been progressively accepted by researchers, following a series of discoveries in various fields of knowledge. The convergence, neither sought nor fabricated, of the results of work that was concluded independently is in itself a significant argument in favor of this theory."

---

* T = *Time Almanac* (2002);
Y = *Yearbook of American and Canadian Churches* (2007);
A = *Major Religions of the World Ranked by Number of Adherents;*
US = reports of the U.S. Census Bureau;
W = Wikipedia.
For a more detailed explanation of these sources, please see appendix note 1 on p. 147.

- 2006. Vatican newspaper *L'Osservatore Romano,* article by Fiorenzo Facchini.

  > "'intelligent design' is not science and teaching it alongside evolutionary theory in school classrooms only creates confusion."

- 2009. Archbishop Gianfranco Ravasi, head of the Pontifical Council for Culture (*The Telegraph* [UK], 11 February 2009). (From Andy Coghlan's article "Vatican Backs Darwin, Dumps Creationism" in the February 11, 2009, *New Scientist* "Short Sharp Science" news blog.)

  > Archbishop Ravasi noted that acceptance of evolution could be traced to St. Augustine and St. Thomas Acquinas, adding "What we mean by evolution is the world as God created it."

Note: The membership figure used in the text, 1.2 billion, is the estimated global figure; for the United States the estimate is approximately 67 million. Fiorenzo Facchini, author of the 2006 article and a professor at the University of Bologna, lamented that certain American creationists had brought the debate back to the "dogmatic" 1800s, and said their arguments were not science but ideology.

**Greek Orthodox Archdiocese of America (1.75 million [$1.5^{Y,A}$–$2.0^{T}$])**

Recent Statements

- 2006. *A Theology of Nature: An Introduction* (first published in 1991), by Metropolitan Paulos Mar Gregorios of the Orthodox Syrian Church of the East, posted on the web site of the Greek Orthodox Archdiocese of America and thus assumed (by me) to be endorsed by the GOAA.

  > "The created order is a space-time process, or rather a procession, orderly and sequential, journeying through life

from something to something. Life is an important aspect of that procession from origin to perfection; it is through the evolution of life that the procession moves forward" (referring to the writings of Gregory of Nyssa in the fourth century).

## EPISCOPAL/ANGLICAN COMMUNION

### Episcopal Church (2.3 million [2.2[Y]–2.4[T]])

Recent statements

- 1981. The 74th Annual Council of the Diocese of Atlanta, in formal action on January 31, 1981, acted without a dissenting vote to oppose by resolution any action by the Georgia Legislature to impose the teaching of Scientific Creationism on the public school system. In an eloquent "Pastoral Statement on Creation and Evolution," the Rt. Rev. Bennett J. Sims, Episcopal Bishop of Atlanta, wrote:

  > "If the world is not God's, the most eloquent or belligerent arguments will not make it so. If it is God's world, and this is the first declaration of our creed, then faith has no fear of anything the world itself reveals to the searching eye of science."

- 1982. Statement from the 67th General Convention of the Episcopal Church.

  > ". . . the House of Bishops concurring, That this 67th General Convention affirm its belief in the glorious ability of God to create in any manner, and in this affirmation reject the rigid dogmatism of the 'Creationist' movement, and be it further resolved, That we affirm our support of the scientists, educators, and theologians in the search for truth in this creation that God has given and entrusted to us."

Additionally, "clergy and scientists from both the Catholic and Evangelical traditions in Anglicanism [= Episcopal Church in the United States] have accepted evolution from Darwin's time to the present. In a resolution passed by the General Convention in 1982, the Church affirmed the ability of God to create in any form and fashion, which would include evolution."

- 2006. Statement from the 75th General Convention of the Episcopal Church.

  "Resolved, That the theory of evolution provides a fruitful and unifying scientific explanation for the emergence of life on earth, that many theological interpretations of origins can readily embrace an evolutionary outlook, and that an acceptance of evolution is entirely compatible with an authentic and living Christian faith . . . " (Resolution A129: Affirm Creation and Evolution).

Note: See "The Episcopal Church: Science, Technology, and Faith" at http://www.episcopalchurch.org/19021_58398_ENG_HTM.htm.

## BAPTIST CHURCHES

### Southern Baptist Convention (16.0 million [15.7[T]–16.3[Y]])

Recent Statements

- 2005. Article in *Time* magazine (August 15, 2005) quoting R. Albert Mohler Jr., president of Southern Baptist Theological Seminary, Louisville, KY. The web site for the issue of the *Baptist Press* with this article is: www.bpnews.net/bpnews.asp?ID=21375

  "Evangelical Christianity and evolution are incompatible beliefs that cannot be held together logically within a distinctly Christian worldview" (view attributed to Mohler in the article; not a direct quote).

- 2009. Mohler was also quoted more recently (February 15, 2009) in the online e-newsletter *Christian Post* (http://www.christianpost.com/article/20090215/evolution-and-christianity-impossible-to-reconcile-says-evangelical-theologian/index.html) as saying that

  > "I find it impossible to reconcile the two" and that "There is no way for God to intervene in the process and for it to remain natural." Yet he is also quoted there as saying that "No Conservative Christian should deny there is a process of change that is evident within the animal kingdom. And there is even a process of natural selection that appears at least to be natural."

**National Baptist Convention, USA (6.6 million [5.0[Y]–8.2[T]])**
Note: The NBC USA was queried in October 2008, but no reply was received. I am assuming their position is similar to the sentiments of the Southern Baptists (see above).

## METHODISM

**United Methodist Church (10.0 million [8.0[Y]–12.0[W]])**

Recent Statements

- 1984. To my knowledge, the UMC does not have an official statement on evolution per se. However, a resolution passed at the 1984 Annual Conference of the UMC in Iowa specifically spoke out against creationism.

  > "*Whereas*, 'Scientific' creationism seeks to prove that natural history conforms absolutely to the Genesis account of origins; and, *Whereas*, adherence to immutable theories is fundamentally antithetical to the nature of science; and, *Whereas*, 'Scientific' creationism seeks covertly to promote

> a particular religious dogma; and, *Whereas*, the promulgation of religious dogma in public schools is contrary to the First amendment to the United States Constitution; therefore, *Be it resolved* that The Iowa Annual Conference opposes efforts to introduce 'Scientific' creationism into the science curriculum of the public schools."

- 2004. The UMC has a 2004 statement on science and technology that would encompass evolution (from the UMC web site at www.umc.org, which in turn comes from *The Book of Discipline of the United Methodist Church*, copyright 2004 by The United Methodist Publishing House).

> "We recognize science as a legitimate interpretation of God's natural world. We affirm the validity of the claims of science in describing the natural world, although we preclude science from making authoritative claims about theological issues. . . . Science and theology are complementary rather than mutually incompatible. We therefore encourage dialogue between the scientific and theological communities and seek the kind of participation that will enable humanity to sustain life on earth and, by God's grace, increase the quality of our common lives together."

- 2008. Resolutions, including a revised wording of the earlier resolution titled "God's Creation and the Church," passed at the 2008 UMC General Conference in Fort Worth, Texas (as reported in the online May 28, 2008, issue of EthicsDaily.com, "Methodists Oppose Teaching of Creationism, Intelligent Design," by Bob Allen, at http://www.ethicsdaily.com/news.php?viewStory=12668).

> Evolution and Intelligent Design (80839-C1-R9999).
> "THEREFORE BE IT RESOLVED that the General Conference

> of the United Methodist Church go on record as opposing the introduction of any faith-based theories such as Creationism or Intelligent Design into the science curriculum of our public schools."
>
> Science and Technology (80050-C1-160.E). ". . . science's descriptions of cosmological, geological, and biological evolution are not in conflict with theology. . . . We find that as science expands human understanding of the natural world, our understanding of the mysteries of God's creation and word are enhanced."

**African Methodist Episcopal Church (AME) (3.8 million [2.5[T,Y]–5.0[W]])**
Note: No reply was received to my query in October 2008. I am assuming that the AME beliefs are aligned with the larger UMC above, as its roots are within the Methodist Church beliefs and doctrines.

## PRESBYTERIANISM

**Presbyterian Church USA (3.3 million [3.0[Y]–3.6[T,US]])**
Recent Statements

- 1998. 85% of Presbyterian pastors responding to a November 1998 PCUSA survey on Science, Technology, and Faith [reported in *The Presbyterian Panel*]) agreed with the statement below.

  > "Evolutionary theory is compatible with the idea of God as Creator."

- 2000. See also "Evolution Not an Option: It Is Essential to Faithfulness" (*SciTech* 9, no. 2 [2000]), by James B. Miller, then-President of the Presbyterian Association on Science, Technology, and the Christian Faith.

- 2002. This statement is from the 214th General Assembly, Columbus, OH.

> "Reaffirms that there is no contradiction between an evolutionary theory of human origins and the doctrine of God as Creator" (Resolution Item 09–08, 2, p. 495).

- Undated Evolution Statement. This statement, reiterating a 1969 statement (GA Minutes 1969: 59-62) by the PCUSA Office of Theology and Worship, is available on the PCUSA web site (www.pcusa.org/theologyandworship/science/evolution.htm).

> "Neither Scripture, our Confession of Faith, nor our Catechisms, teach the Creation of man by the direct and immediate acts of God so as to exclude the possibility of evolution as a scientific theory."

Note: The PCUSA should not be confused with the more conservative Presbyterian Church in America (PCA), currently estimated at 300,000 to 500,000 members, mostly in the southeastern United States and especially in Atlanta. The PCA does have a "Report of the Creation Study Committee" available on their web site that indicates that they could not reach consensus on this subject.

## CONGREGATIONALISM

### United Church of Christ (1.3 million [1.2$^{Y}$–1.4$^{T}$])

Recent Statements

- 1992. United Church Board for Homeland Ministries (statement titled "Creationism, the Church, and the Public School") at: http://ncseweb.org/media/voices/united-church-board-homeland-ministries.

> "We acknowledge modern evolutionary theory as the best present-day scientific explanation of the existence of life on earth; such a conviction is in no way at odds with our belief in a Creator God, or in the revelation and presence of that God in Jesus Christ and the Holy Spirit."

- 2008. Rev. John H. Thomas, General Minister and President, United Church of Christ, "A New Voice Arising: A Pastoral Letter on Faith Engaging Science and Technology" (January 2008) (www.ucc.org/not-mutually-exclusive/pdfs/pastoral-letter.pdf).

  > "Evolution helps us see our faithful God in a new way. Evolution also helps us see ourselves anew, as creatures who share a common origin with other species."

## LUTHERANISM

### Evangelical Lutheran Church in America (5.0 million [4.8[Y]–5.2[T,US]])

Recent Statements

- Lutheran World Federation, statement on the web at www.lutheranworld.org/News/LWI/EN/1823.EN.html. See also Edwin A. Schick, "Evolution" in *The Encyclopedia of the Lutheran Church,* vol. 1, ed. J. Bodensieck (Minneapolis, MN: Augsburg Publishing House, 1965).

  > "In whatever way the process may be ultimately explained, it has come about that an idea which has been most thoroughly explored in the field of biology (lower forms of life evolving into higher) has by means of organismic analogy found universal application. Phenomena thus accounted for range from physical realities (evolution of the atoms and expanding galaxies) to man and his social experience (the evolution of cultural values) including his understanding of time and history (the evolutionary vision of scientific eschatology)."

Note: The Lutheran World Federation statement should not be taken as representative of all Lutheran congregations. The largest Lutheran group in the United States is the Evangelical Lutheran Church in America (ELCA), which, with 5 million members, is also the fifth largest "mainline" or "historical" Protestant denomination in the nation. On the ELCA web site under the heading "Creation vs. Evolution," the following statement appears:

> "The ELCA does not have an official position on creation vs. evolution, but we subscribe to the historical-critical method of biblical interpretation, so we believe God created the universe and all that is therein, only not necessarily in six 24-hour days, and that God actually may have used evolution in the process of creation. In fact, to deny the possibility that evolutionary processes were used is seen by some as an attempt to limit God's power. 'Historical criticism' is an understanding that the Bible must be understood in the cultural context of the times in which it was written."

**Lutheran Church—Missouri Synod (2.5 million [2.4[Y]–2.6[T]])**
Note: In stark contrast to the ELCA, the Lutheran Church—Missouri Synod is strongly against the idea of compatibility of evolution and faith; their web site formerly referred readers to a 1977 document favoring strict literal creationism and now includes a more recent pro-ID statement, "What About Creation and Evolution," by former LCMS president Dr. A. L. Barry (www.lcms.org/graphics/assets/media/LCMS/wa_creation-evolution.pdf). The third largest Lutheran group in the United States, the Wisconsin Evangelical Lutheran Synod (~ 390,000), is not treated here.

## ANABAPTISTS

**Combined Anabaptist Groups (4.5 million [4.0–5.0[W]])**
Note: The American religious groups often combined under the heading "Anabaptist" (including Mennonites, Brethren in Christ, Amish, Hutterites, Church of the Brethren, Open Brethren, and sometimes the Religious Society of Friends [Quakers]) are primarily concerned with how

Christians should address the many social ills facing humanity and bring about peace in the world. As such, they have no stated position on science, faith, evolution, and creation—although I have corresponded with individual members who do have a position on the issue—and therefore I am unable to attribute to these groups one particular stance or the other. Some classifications of religious organizations include Baptists (which I have treated separately above) among the Anabaptist groups.

## PENTECOSTALISM

### Assemblies of God, USA (2.65 million [2.5[T,US]–2.8[Y]])

Recent Statements

- 1977. The Assemblies of God USA web site (at http://ag.org/top/Beliefs/Position_Papers/index.cfm) lists a number of position papers including "The Doctrine of Creation," adopted by the Assemblies of God General Presbytery in August 1977 and containing the statement below. Another article in their section on beliefs is titled "Creationism" and was endorsed by the church's Commission on Doctrinal Purity and the Executive Presbytery; it states that "Assemblies of God believers hold that the Genesis account should be taken literally."

    > "The Bible record of creation thus rules out the evolutionary philosophy which states that all forms of life have come into being by gradual, progressive evolution carried on by resident forces. It also rules out any evolutionary origin for the human race, since no theory of evolution, including theistic evolution, can explain the origin of the male before the female, nor could it explain how a man could evolve into a woman."

**Church of God (Cleveland) (2.9 million [0.8[T,US]–5.0[W]])**

Recent statements

- 1980. Resolution on Creationism (on their web site www.churchofgod.org/ under "Beliefs/Resources").

> "WHEREAS secular humanism and anti-God philosophies are being taught in our public educational systems; and WHEREAS there is a need for God's people to unite against the teaching of evolution as a scientific fact; THEREFORE BE IT RESOLVED that we give our full support to the principle that where evolution is taught in our public schools, provision be made for teaching the Biblical alternative of creation."

Note: This refers only to the Church of God with international offices in Cleveland, Tennessee. Estimates of membership vary widely.

**Church of God in Christ (5.5 million [5.5[Y]])**

Notes: Although the COGIC separated from the Assemblies of God, its beliefs remain firmly Pentecostal and in line with the AOG; it is therefore listed as not being accepting of evolution as compatible with its faith in table 1.

**International Circle of Faith (5.5 million[W])**

Recent Statements

- 2008. Dr. Bernie L. Wade, Chancellor, International Circle of Faith Colleges and Seminaries (personal communication, December 26, 2008 e-mail).

> "The position of the International Circle of Faith concerning creation is biblical creationism. Biblical Creationism is the belief that the biblical record of creation is true. The biblical creationist believes that God created the entire universe out of nothing, by the power of His word. . . . We believe the first chapter of Genesis describes real

events and that the world underwent growth and development during its creation. Further, the human race is unique, the crown of creation. This includes a historical Adam and Eve and a historical fall from grace."

Note: The only estimate I found for ICOF membership was 10.9 million worldwide; I have not received a membership figure for within the United States, and so I am using half of that figure as my estimate here. The sole source was far too high on other estimates; the same may be true here. Other Pentecostal churches possibly with large memberships, including the New Apostolic Church and the International Church of the Foursquare Gospel, did not respond to my queries, and no statements could be found. It seems unlikely that any of these would be generally supportive of evolution, given that all are Pentecostal congregations and thus assumed (by me) to be more similar to the Assemblies of God, Church of God, and ICOF positions above. Thus, ICOF is listed as not being accepting of evolution as compatible with its faith in table 1, but without membership estimates.

## RESTORATIONISM

### Seventh-Day Adventists (0.9 million [0.8[T,US]–15.0[W]])

Recent Statements

- 2004. The International Faith & Science Conferences 2002–2004 report of the Organizing Committee to the General Conference Executive Committee through the office of the General Conference President, September 10, 2004. From the "Fundamental Beliefs" section of the Seventh-Day Adventists web site (www.adventist.org/).

  "We strongly endorse the document's affirmation of our historic, biblical position of belief in a literal, recent, six-day

> Creation" (from their response statement to "An Affirmation of Creation" on the above web site).

Note: Membership estimates ranged extremely widely, from 0.8 to 15 million, for this group. However, most sources used 0.8 to 0.9 million; I am using the higher of these two figures. Seventh-Day Adventists have a long history as confirmed young-earth creationists; members were responsible for some of the books on young-earth geology early in the twentieth century.

### Church of Christ (3.3 million [1.5[US,Y] – 5.0[W]])

Note: This treatment is restricted to the "restorationist" Church of Christ group in America, as opposed to more than twenty other groups that share "Church of Christ" as part of their name (see, for example, the Wikipedia entry on "Church of Christ"). A difficult group to define, the Church of Christ movement can also include churches by the name of "Disciples of Christ" and simply "Christian Church." Individual churches are autonomous, not denominational, and thus there is no central organizing body. The web site "Churches of Christ Online" (at http://cconline.faithsite.com/) did not have a science/faith statement on it as of August 2009. Links provided there lead me to think that the overall Church of Christ position does not embrace evolution as compatible with their beliefs; this might be incorrect, and so I have placed them in the "unclear" category in table 1.

## NON-DENOMINATIONAL EVANGELICALISM

### Calvary Chapel (3.3 million[5])

Note: Calvary Chapel and the Vineyard (*below*) are best described as non-denominational church movements or fellowships. Individual congregations have some flexibility in their stated beliefs and practices. Because of this flexible structure, the primary web site for Calvary Chapel states that there is "no central headquarters or organization," and the reader is directed to the nearest Calvary church. I queried the original Calvary Chapel in Costa Mesa, California, but did not receive a reply. The main web site (at www4.calvarychapel.com/) states that Calvary Chapel is "focused on the inerrancy of the Bible;" Calvary is also described as a "Bible-believing, evangelical church" (from the "Calvary Chapel Distinctives" written by founding pastor Chuck Smith). These statements, and Pastor Smith's notes on the Calvary web site asking why "natural selection today creates a deterioration and mongrelization of a species, rather than a highly developed form," indicate a leaning away from seeing evolution as part of God's creation. In 2009 several Calvary Chapels hosted anti-evolution workshops or conferences associated with Ken Ham's creationist "Answers in Genesis" movement. Thus, I have listed Calvary in table 1 as being opposed to the idea of compatibility between evolution and their faith.

**The Vineyard (0.2 million)**

Note: The Vineyard, which broke away from the Calvary movement (*above*) in the early 1980s, is similar in allowing a diversity of worship styles and approaches, but it is perhaps more diverse and less constrained than Calvary. Within the Vineyard one can find a wide range of stances on the topic of evolution and creation, from liberal or progressive advocates who see evolution as an act of and expression of God to more evangelical advocates who would not embrace that view (personal communication, Kris Miller, The Vineyard USA National Office, October 2008). For these reasons, I have listed The Vineyard movement as "unclear" in table 1. The membership estimate (150,000 to 200,000) was given to me by Pam Trautmann of the Vineyard National Office (personal communication, April 2009).

**Christian "Megachurches"**

Note: A relatively recent phenomenon in Protestant Christianity is the rise of extremely large, usually nondenominational churches, particularly in the United States. The term "megachurch" has sometimes been applied to those that see 2,000 or more attendees per week. These churches tend to be evangelical or Pentecostal, and they may be independent of mainline Protestant denominations that are similar in name. Because these churches probably constitute a significant (but currently unknown) percentage of the U.S. population of Christians, some mention should be made of them here. A list of the top 100 churches (ranked by size) is published each fall by *Outreach* magazine (http://www.outreachmagazine.com/top_100.asp). I queried each church that appeared in the 2008 top 10 list about the compatibility of their faith with evolutionary biology, and the results are below. The name of the church is followed by the number of weekly attendees and their position (if known) based on their reply or on known affiliations.

1. Lakewood Church (Houston), 43,500 (No reply)
2. Second Baptist (Houston), 23,659, No (affiliated with Southern Baptist Convention)
3. North Point Community Church (Alpharetta, Georgia), 22,557, Unknown (e-mail indicating a broad acceptance of, and appreciation for, diversity)
4. Willow Creek Community Church (Illinois), 22,500 (No reply)
5. Life Church (Edmond, Oklahoma), 20,823, No ("We believe in creation plain and simple;" e-mail from Randy Coleman, August 24, 2009)
6. West Angeles Cathedral (Los Angeles), 20,000, No (affiliated with Church of God in Christ)
7. Fellowship Church (Grapevine, Texas), 19,913 (No reply)
8. Saddleback Church (Lake Forest, California), 19,414, No (affiliated with Southern Baptist Convention)
9. Calvary Chapel Fort Lauderdale (Florida), 18,000, No (affiliated with Calvary Chapels)
10. Potter's House (Dallas), 17,000, No (described as a fundamentalist Pentecostal Church)

The above numbers are not included in table 1. Some of these churches, such as those affiliated with the Southern Baptist Convention (Second Baptist, Saddleback) or with Calvary Chapel (Calvary Chapel

Fort Lauderdale) may have been included in the membership estimates of mainstream denominations given earlier. There is also a lot of volatility in the membership of these large churches; the number 6 and 7 churches on the 2008 list did not make the top 10 in 2007. Because most of these churches, and indeed most of the top 100 on the *Outreach* list, would fit most comfortably in the "No" column of table 1, these churches would increase the total for that category.

# Notes

INTRODUCTION

Epigraph. Unless otherwise noted, biblical quotations are from the *New Revised Standard Version, The HarperCollins Study Bible, Including the Apocryphal/Deuterocanonical Books with Concordance*, edited by Harold W. Attridge (New York: Harper One Publishers, 2006).

1. A November 1998 PCUSA survey on Science, Technology, and Faith (reported in *The Presbyterian Panel*) found the statement "evolutionary theory is compatible with the idea of God as Creator" was agreed to by 61% of church members and 85% of Presbyterian pastors. See also "Evolution Not an Option: It Is Essential to Faithfulness" (*SciTech* 9, no. 2. [2000]), by James B. Miller, then-President of the Presbyterian Association on Science, Technology, and the Christian Faith. See also the PCUSA statements in the appendix.
2. For a concise review of some of the recent literature, see Glenn Branch's excellent article "Understanding Creationism after *Kitzmiller*" (*BioScience* 57, no. 3 [2007]: 278–84).
3. My degrees are a BS in Zoology, University of Kentucky, 1978; MS in Biology, University of Southwestern Louisiana (now the University of Louisiana, Lafayette), 1981; and PhD in Biological Sciences, Florida State University, 1986.
4. For polls indicating that a large percentage of scientists are people of faith, see L. Witham, "Many Scientists See God's Hand in Evolution," *Washington Times*, April 11, 1997, A8, reprinted on the National Center for Science Education (NCSE) web site, and also E. J. Larson and L. Witham, "Scientists Are Still Keeping the Faith," *Nature* 386 (1997): 435–36. According to a survey of members of the American Association for the Advancement of Science by the Pew Research Center in 2009, 51% said they believe in God or a higher power (article by David Masei, "Faith Inside Science," *Los Angeles Times*, November 24, 2009). In a 2005 informal "shovelbums" survey, more than 7,700 scientists in only four days signed a petition in support of the continued teaching of evolution in science classrooms. Of those responding, only 36% indicated that they were agnostic or atheistic, with 64% either indicating that they were people of faith (including Buddhists, Hindus, Jews, and members of fifteen Christian denominations) or not responding to that question (Mark Siddall, personal communication, American Museum of Natural History; see "Thousands of Scientists Sign Petition Opposing the Teaching of Intelligent Design as Science,"

www.prnewswire.com/cgi-bin/stories.pl?ACCT=104&STORY=/www/story/10-20-2005/0004173856&EDATE).

See the appendix for a breakdown of the position of various denominations—including statements made by various religious leaders that lend support to the idea that evolution is compatible with faith. Also see the NCSE web site, especially their "Voices for Evolution" document, and the Clergy Letter Project (www.butler.edu/clergyproject/rel_evol_sun.htm), which has many links and other resources detailing the efforts of religious leaders and scientists toward this end. As of September 1, 2009, 11,951 Christian clergy had signed the letter endorsing the continued teaching of evolution; 457 rabbis and 205 Unitarian Universalists signed similar letters.

5. An exception I would make here is that occasionally science is used to argue against some specific religious claims, such as the notion that a Sun God pulls the sun across the sky in a chariot (ancient Greece) or that the sun revolves around the earth, as was once believed to be a biblically based truth.
6. Peter Gomes, *The Good Book: Reading the Bible with Mind and Heart* (New York: William Morrow and Company, Inc., 1996), p. 316.
7. Perhaps this "middle ground" is best expressed by such writers as Arthur Peacocke, a physical biochemist and Anglican priest: "As a theist—one who considers that the best explanation of the existence and lawfulness of the natural world is that it depends for its existence and inbuilt rationality on a self-existent Ultimate Reality (a Creator 'God')—I find the epic of evolution, from the 'Hot Big Bang' to *Homo sapiens*, an illumination of how . . . God is and has been creating. Evolution enriches our insights into the nature and purposes of the divine creation. . . . I regard God as creating in, with, and through the natural as unveiled by the sciences; hence I espouse a 'theistic naturalism.'" (www.pbs.org/wgbh/evolution/religion/faith/statement_03.html).

## CHAPTER 1. BY THE NUMBERS

Epigraph. Although this latter quote is often attributed to Albert Einstein, it probably did not originate with him (R. Keyes, *The Quote Verifier* [New York: St. Martin's Griffin, 2006], p. 54, citing research by Alice Calaprice in *The New Quotable Einstein*).

## CHAPTER 2. THE PRISM AND THE RAINBOW

Epigraph. Goethe was an incredibly influential eighteenth-century German writer, philosopher, and polymath. Ironically, he thought Newton's ideas on colors and prisms were incorrect.

1. Although the original word used is *bow,* there is no question about what is being discussed, so many modern translations employ the word *rainbow.*
2. That scientific knowledge of how a rainbow works should be an inspiration, rather

than otherwise, was also the source of the title for Richard Dawkins's book *Unweaving the Rainbow: Science, Delusion, and the Appetite for Wonder* (New York: Mariner Books, 2006). In Dawkins's words, "My title is from Keats, who believed that Newton had destroyed all the poetry of the rainbow by reducing it to the prismatic colours. Keats could hardly have been more wrong, and my aim is to guide all who are tempted by a similar view towards the opposite conclusion. Science is, or ought to be, the inspiration for great poetry . . ." (p. x). Dawkins's sentiment here is one with which I wholeheartedly agree.

## CHAPTER 3. THE FLAT EARTH SOCIETY

1. Educated ancient Greeks certainly knew that the world was round, and Eratosthenes, director of the great library of Alexandria, Egypt, even calculated its circumference to a surprising degree of accuracy (within 50 miles of our current figures) using trigonometry. For a thorough review of the history of belief in the concept of a flat earth, see Christine Garwood, *Flat Earth: The History of an Infamous Idea* (New York: MacMillan, 2007).
2. I am not sure who is in charge of the Flat Earth Society today, or even if it still exists. If you Google the society, you'll run into quite a few web sites. Some of them, like "The Flat Earth Society: Deprogramming the masses since 1547," appear to be run by some people just having some fun. That is, I don't think that the authors of this site mean much (if any) harm, and their arguments are as inventive as they are amusing (at one point they warn you that your dog is already a member). At least on this web site they do not refer to any biblical passages in support of the earth being flat.

   Information on the "real" Flat Earth Society—a society that earnestly believed in and promoted a flat-earth view of the world—is summarized in Christine Garwood's (2007) book (see note 1). The president of the Flat Earth Society in the United States was a man named Charles K. Johnson, an incredibly colorful and prolific (though grammatically challenged) writer. Under Johnson's presidency for more than three decades, the Flat Earth Society grew to about 3,000 members worldwide, published a newsletter called "Flat Earth News," and was active as recently as 1995, when Johnson's home outside Lancaster burned to the ground. Johnson died in March 2001.

   In September 2008, an online article on FoxNews.com ("Believers in Flat Earth Not about to Change Minds") referred to a BBC report that the Flat Earth Society still existed, quoting "Tennessee-based computer scientist and society member John Davis" and James McIntyre (a Briton) in defense of the Society. Are they serious? I really don't know.
3. The father of the modern Flat Earth Society was an Englishman named Samuel Birley Rowbotham, who sometimes wrote and spoke under the pen name "Parallax" and who authored three editions of a lengthy book called *Zetetic Astronomy: Earth Not a Globe*, the last chapter of which cited 76 Bible verses in support of a flat earth. Lady Elizabeth Blount, founder of a flat-earth advocacy group called the Universal

Zetetic Society, a forerunner of the modern Flat Earth Society, was adamant about biblical support of a flat and immovable earth. Garwood (2007) quotes Blount as writing "We cannot divorce the religion of the Bible from the science of the Bible, hence the globists [round-earth believers] cannot be Christians—nor can Bible Christians be followers of Newton's philosophy" (p. 157).

4. Some of the many verses cited to defend belief in a "fixed" (immovable) earth are:

    1 Chronicles 16:30: "The world is firmly established; it shall never be moved"

    Psalm 93:1: "He has established the world; it shall never be moved"

    Psalm 96:10: "The world is firmly established; it shall never be moved"

    Psalm 104:5: "You set the earth on its foundations, so that it shall never be shaken"

5. A number of evangelical Christians today, and especially the modern creationists (those advocating the teaching of intelligent design [ID] in our schools), are somewhat embarrassed about any association of flat-earth thinking with Christianity. That any "erroneous" belief was once thought to be scripturally based is seen as an insult, and so they go to great lengths to deny that Christians, or anyone at all, ever seriously thought the world was flat. As pointed out earlier, they are largely correct; the view that a flat-earth mentality was widespread during the Middle Ages and was shared by large numbers of Christians is a myth. However, the more modern example (the Flat Earth Society headquartered in California, which traces its roots to the Zetetic writings of Samuel Rowbotham, Elizabeth Blount, Samuel Shenton, Wilbur Glenn Voliva, and others) is, or at least was, quite real. Some modern creationists and most ID proponents readily acknowledge that the earth is not flat and that it is not young, yet still try hard to deny the abundant evidence for, and general acceptance of, evolution and other sciences (see chapters 9 and 10).
6. St. Augustine's book on the difficulties in interpreting Genesis 1 and 2 literally was titled *On the Literal Interpretation of Genesis* (translated and annotated by John Hammond Taylor, S. J. [New York: Newman Press, 1982]). An article describing Augustine's book and its relevance to today's Christians entitled "The Contemporary Relevance of Augustine's View of Creation" can be found on the American Scientific Affiliation web site (www.asa3.org/ASA/PSCF.html). See also "Literalism" on the Talk Origins web site at: www.talkorigins.org/indexcc/CH/CH102.html.

## CHAPTER 4. OF SERPENTS AND CERTAINTY

1. For more on the history of snake handlers, see Thomas G. Burton, *Serpent Handling Believers* (Knoxville: University of Tennessee Press, 1993) and David L. Kimbrough,

*Taking Up Serpents: Snake Handlers of Eastern Kentucky* (Macon, GA: Mercer University Press, 2002).

2. In a radio interview with Jolo churchgoers, David Isay interviewed several church members and asked their reasons for their unusual behavior. Isay recorded one church member as saying, "If man could have written that part of God's word, he could have written any of it. And if I doubted that part of God's word, then it'd give me a reason to doubt the rest of it." Another congregant said simply, "A lot of people say 'Well, you interpret it wrong.' I hadn't got but one way to interpret it, and that's read it for what it says." Dewey Chafin, one of the better-known snake handlers (whose sister died at age 22 from a snake bite), responded to Isay's direct question with an equally direct reply: "Why I do it is because the Bible said to do it. And that's just the bottom line." (Isay, "They Shall Take Up Serpents," *All Things Considered,* November 30, 1992.)
3. Several respected biblical scholars are of the opinion that these verses (Mark 16:9–20)—frequently referred to as the "longer ending" of Mark—were not part of the original gospel, as the vocabulary, literary style, and content are different from earlier passages of Mark and because this ending is missing from the earliest manuscripts of Mark that we possess. According to Douglas R. A. Hare (*Mark*, Westminster Bible Companion [Louisville, KY, Westminster John Knox Press, 1996]), the existence of two different endings to the Gospel of Mark "indicates that neither is original. Each was added because scribes found Mark's conclusion in 16:8 too abrupt" (p. 227). James Edwards (*The Gospel According to Mark*, The Pillar New Testament Commentary [Grand Rapids, MI: Wm. B. Eerdmans Publishing Company, 2002]) is more forceful in his analysis: "It is virtually certain that 16:9–20 is a later addition and not the original ending of the Gospel of Mark" (p. 497). Both Hare and Edwards point out, as have others, that there also are no parallels to this story in any of the other Gospels. See also J. R. Donaghue, S.J., and D. J. Harrington, S.J., *Sacra Pagina,* vol. 2: *The Gospel According to Mark* (Collegeville, MN: The Liturgical Press, 2002).
4. Exposing yourself to danger intentionally (either by snakebite or poison) would seem to be at odds with other statements attributed to Christ, such as "Do not put the Lord your God to the test" (Matthew 4:7). And indeed, most Christians do not handle venomous snakes as a sign of faith, possibly because they understand that they can use the intelligence God has given them to find deeper meanings in the Scriptures.

## CHAPTER 5. THE NATURE OF SCIENCE

1. The potential to test an idea, and the ability to repeat that test, with tests and results shared among researchers, cannot be overstressed. Science is perhaps best defined by the idea of a testable, repeatable question such that unrelated workers can perform the same test and expect the same results.

   Some critics have expressed a concern about science being "dogmatic," when nothing could be further from the truth. Science is by definition testable and subject to revision and change. Moti Ben-Ari, in his book *Just a Theory: Exploring the Nature of Science* (New York: Prometheus Books, 2005), notes that "if it might be wrong, it's science" (p. 63) (though non-science claims can also be wrong, of course). This differs from how we tend to view religious beliefs, which in general are not open for revision and are not testable, or at least not testable in the same way.

2. There is also a growing awareness that the "scientific method" as usually described is not an accurate reflection of how science operates; it is too linear and does not adequately explain the more complex, interactive, and non-linear processes involved. Visit www.understandingscience.org to read a more accurate depiction of how science operates.
3. Similarly, the *limitation* of science also has to do with finding explanations that can be tested in a particular way. Science requires tests that can be shared among researchers. Faith has other means of "testing," including personal experience and personal revelation, which science does not have. Although originally many of the natural sciences were aligned with religion, scientists eventually came to exclude explanations that could not be tested by shared experience or experimentation.
4. In the original sense of the word, an evangelist was simply someone spreading the Christian faith, either by word or by example. By that definition, all Christians could be considered evangelicals, because all are called upon to follow the teachings and the example set by Jesus Christ (following the Great Commission in Matthew 28:16–20 and elsewhere). In recent decades there has been a growing awareness of a group of outspoken and socially conservative Christians, mostly in the United States, whose beliefs and practices differ from those of mainstream Catholic and Protestant denominations to the extent that a separate descriptive term was needed. None has been proposed, although "neo-evangelical" was used by Ted V. Foote and P. Alex Thornburg (*Being Presbyterian in the Bible Belt: A Theological Survival Guide for Youth, Parents, and Other Confused Presbyterians* [Louisville, KY: Geneva Press, 2000], pp. xiii and 81), and the word evangelical is often used in a narrow sense for this group. These Christians usually have in common a strong belief in the need for a personal conversion (being "born again"), an imperative to evangelize or proselytize about their faith, and a high regard for biblical authority (often including the insistence that the Bible is true historically and scientifically). For more information on the uses and misuses of the term evangelical, see the web site of the Ontario Consultants on Religious Tolerance at: www.religioustolerance.org/evan_defn.htm.
5. Most of Galileo's assertions and observations about our universe had been accepted by the Church long before that time, and the public at large certainly was aware that the earth circled the sun rather than vice versa. The declaration by Pope John Paul in 1992, after thirteen years of deliberation by a commission of investigators established in 1979, was more of an official statement of Galileo's vindication, but it still serves to make the point that scientific discoveries are not always accepted overnight. For more on the Galileo trial and its influence through history see M. A. Finocchiaro, *Retrying Galileo, 1633–1992* (Berkeley: University of California Press, 2005).

## CHAPTER 6. WHAT DOES "THEORY" MEAN?

1. For a discussion of the use and misuse of the word "theory," see Moti Ben-Ari's book, *Just a Theory: Exploring the Nature of Science* (New York: Prometheus Books, 2005).
2. Here's an interesting historical comparison to such comments about evolution today. In April 1615, Cardinal Roberto Bellarmine, responding to the published writings of Galileo, cautioned scientists to treat Copernican views, where the earth orbited the sun instead of vice versa, as hypothesis, not fact. Because he was no astronomer, we can only deduce that his purpose was to infuse Galileo's findings with doubt, much as is attempted today when evolution is referred to as "just a theory."
3. In this definition of "fact" I follow mostly the arguments in K. Fitzhugh, "Fact, theory, test, and evolution," *Zoologica Scripta* 37 (2007): 109–13 (quote from p. 109). See also T. R. Gregory, "Evolution as fact, theory, and path," *Evolution: Education and Outreach* 1 (2008): 46–52.
4. National Academy of Sciences of the United States, *Teaching about Evolution and the Nature of Science* (Washington, D.C.: National Academies Press, 1998), p. 5.
5. Kenneth R. Miller, *Finding Darwin's God: A Scientist's Search for Common Ground between God and Evolution* (New York: Cliff Street Books/Harper Collins, 1999).
6. Newton, himself, did not really have a stated theory of how gravity worked, just the observation that there was a constant relationship between mass and force.
7. For a good discussion of other ways to think about gravity (one that I found very disquieting), see Juan Maldacena's article "The Illusion of Gravity," first published in *Scientific American* and reprinted in B. Greene, ed., *The Best American Science and Nature Writing* 2006 (Boston: Houghton Mifflin Company, 2006).

## CHAPTER 7. WHAT IS EVOLUTION?

1. Bruce Albert, past president of the National Academy of Sciences, in his preface to the NAS booklet on science and creationism (*Science and Creationism: A View from the National Academy of Sciences,* second edition [Washington, DC: National Academy Press, 1999]), writes

> the concept of biological evolution is one of the most important ideas ever generated by the application of scientific methods to the natural world. The evolution of all the organisms that live on Earth today from ancestors that lived in the past is at the core of genetics, biochemistry, neurobiology, physiology, ecology, and other biological disciplines. It helps to explain the emergence of new infectious diseases, the development of antibiotic resistance in bacteria, the agricultural relationships among wild and domestic plants and animals, the composition of Earth's atmosphere, the molecular machinery of the cell, the similarities between human beings and other primates, and countless other features of the biological and physical world. (pp. viii–ix)

2. When I say "over time," it is important to realize that, although some evolutionary change can be extremely rapid, sometimes this means a truly vast amount of time, a period that is difficult to imagine by the standards of our all-too-brief lives. This immense amount of time is what allows a population to slowly change through the gradual accumulation of small genetic changes, for it is not an individual that changes over time, but a population.
3. See Keith Thomson, *Before Darwin: Reconciling God and Nature* (New Haven, CT: Yale University Press, 2007).
4. Jerry A. Coyne, *Why Evolution Is True* (New York: Viking Press, 2009).
5. There are a handful of PhD-level scientists and philosophers who have written articles or books arguing against the accepted findings of evolutionary biology. The names that most often appear in this vein are those of biochemist Michael Behe and philosopher William Dembski. But they have not been active in publishing mainline, reviewed articles that demonstrate or uphold their beliefs in respected scientific journals, and they are at odds with the huge number of active, publishing scientists working today. It's also good to keep in mind that writing a book with the word "science" in the title does not make one a practicing scientist.
6. Because 2009 was the 200th anniversary of the birth of Charles Darwin, several prominent magazines, including *Scientific American, Nature,* and *National Geographic*, devoted issues to examples of evolution being studied today, as did an earlier article in *Natural History* (114, no. 9 [November 2005]). These, plus a recent spate of books on the topic (e.g., *Your Inner Fish: A Journey into the 3.5-Billion-Year History of the Human Body*, by Neil Shubin [New York: Pantheon, 2008]; *Why Evolution Is True*, by Jerry Coyne [New York: Viking, 2009]; *The Tangled Bank: An Introduction to Evolution*, by Carl Zimmer [Greenwood Village, CO: Roberts and Company, 2009]; *The Greatest Show on Earth: The Evidence for Evolution*, by Richard Dawkins (New York: Free Press, 2009]; and *Evolution: What the Fossils Say and Why It Matters,* by Don Prothero [New York: Columbia University Press, 2007]) serve as good introductions to some of the many lines of supportive evidence.
7. For statements in support of evolution from the Vatican, see the appendix. Billy Graham's feelings about the compatibility of evolutionary theory with theology are perhaps most clearly voiced in the book *Billy Graham: Personal Thoughts of a Public Man,* by David Frost and Fred Bauer (Colorado Springs, CO: Chariot Victor Publishing, 1999), where he is quoted as saying:

> I don't think that there's any conflict at all between science today and the Scriptures. I think we have misinterpreted the Scriptures many times and we've tried to make the Scriptures say things they weren't meant to say. I think that we have made a mistake by thinking the Bible is a scientific book. The Bible is not a book of science. The Bible is a book of Redemption, and of course I accept the Creation story. I believe that God did create the universe. I believe that God

> created man, and whether it came by an evolutionary process and at a certain point He took this person or being and made him a living soul or not, does not change the fact that God did create man . . . whichever way God did it makes no difference as to what man is and man's relationship to God. (p. 72)

8. In 2005, thirty-eight winners of the Nobel Prize signed a petition endorsing the continued teaching of evolution in the world's classrooms.
9. Nancey Murphy, "Nature's God: Nancey Murphy on Religion and Science" (interview), *The Christian Century* (December 2005): 20–26. Quote from p. 23.

## CHAPTER 8. WHAT IS CREATIONISM?

1. See "progressive creationism" in the glossary; see also Eugenie Scott's book *Evolution vs. Creationism: An Introduction*, 2nd ed. (Westport, CT: Greenwood Press, 2008) for a discussion of the full range of different creationist views of the world, and www.ncseweb.org/creationism/general/creationevolution-continuum.
2. Ken Ham's Creation Museum in northern Kentucky, which promotes a young earth, is the most recent and most notable of these.
3. Some of the more outlandish earlier claims that creationists made in print were reviewed in a small volume edited by Liz Rank Hughes (*Reviews of Creationist Books*, second edition [Berkeley, CA: The National Center for Science Education, 1992]).
4. It is no coincidence that the state of Kansas was also singled out in 2000 as having the nation's worst science standards for public schools, according to a national education survey conducted by the Thomas B. Fordham Institute. According to the report based on the survey ("Good Science, Bad Science: Teaching Evolution in the States") Kansas "now makes a mockery of the very definition of science." Kansas was not alone; fourteen other states also received a grade of F whereas only seven received an A for their science teaching standards. By 2005 there had been very little improvement ("The State of State Science Standards," Thomas B. Fordham Institute, 2005). The Kansas standards were corrected in 2007.
5. The wording of the stickers placed inside biology textbooks in Cobb County read: "This textbook contains material on evolution. Evolution is a theory, not a fact, about the origin of living things. This material should be approached with an open mind, studied carefully, and critically considered." After the court ruled against them, in a response typical of religiously motivated school boards, the disappointed Cobb County Board of Education replied that "textbook stickers are a reasonable and even-handed guide to science instruction." One has to wonder, if this is indeed true, who would be allowed to write the content of these stickers and to decide which of our textbooks should receive them.
6. See *The Pillars of Creationism*, on the web site of the NCSE: http://ncseweb.org/.

7. William Thwaites, in "Brief History of Creationism—From Middle Ages to 'Creation Science,'" on the NCSE web site at www.ncseweb.org/creationism/general/brief-history-creationism.
8. "Evolution and Creationism in Public Education: An In-Depth Reading of Public Opinion," report of a poll conducted in 1999 (published in 2000) by DYG, Inc., on behalf of the People for the American Way Foundation and found in the Report Archives section of the PFAW web site at http://67.192.238.59/multimedia/pdf/Reports/evolutionandcreationisminpubliceducation.pdf (last accessed Sept. 9, 2009). The number 29% in the text comes from 13% ("support treating evolution and creationism equally") and 16% ("support creationism-oriented positions") on page 15 of the report.
9. Kenneth R. Miller, *Finding Darwin's God: A Scientist's Search for Common Ground between God and Evolution* (New York: Cliff Street Books/Harper Collins, 1999), p. 267.
10. Dietrich Bonhoeffer, *Letters and Papers from Prison: The Enlarged Edition,* edited by Eberhard Bethge (New York: MacMillan, 1972), pp. 311–12.
11. The arguments of creationists have been covered in detail in the books by Kenneth Miller, Massimo Pigliucci, Michael Ruse, Barbara Forrest, and Eugenie Scott. See the recommended reading at the back of this book for publication details.
12. Literal and inerrant creationists also run into a problem within Genesis in that there are two creation stories there—in Genesis 1 and 2—with different names for God, different orders of creation, and different literary genres (poetry vs. prose); see chapter 12. As with earlier examples, the profound messages of Genesis are not lost by the knowledge that there are two accounts here, whereas a strictly literal reading causes these messages to become muddled in the attempt to reconcile these different stories.
13. Rev. James W. Skehan, "The Age of the Earth, of Life, and of Mankind: Geology and Biblical Theology versus Creationism," in R. W. Hanson, ed., *Science and Creation: Geological, Theological and Educational Perspectives* (New York: MacMillan Publishing Company, 1986), pp. 10–32.

## CHAPTER 9. WHAT IS INTELLIGENT DESIGN?

1. For a complete chronology see Eugenie Scott, *Evolution vs. Creationism: An Introduction,* second ed. (Westport, CT: Greenwood Press, 2008); the article by Nick Matzke, "But Isn't It Creationism? The Beginnings of 'Intelligent Design' in the Midst of the *Arkansas* and *Louisiana* Litigation" in *But Is It Science? The Philosophical Question in the Creation/Evolution Controversy,* updated edition, edited by Robert Pennock and Michael Ruse (New York: Prometheus Books, 2009); and the decision in the Dover, Pennsylvania, trial (*Tammy Kitzmiller et al. v Dover Area School District,* Case 4:04-cv-02688-JEJ, Document 342, filed 12/20/2005, on the web at www.ncseweb.org/webfm_send/73).

2. In fact, one of the books written by creationists for use as an alternative text for evolutionary science is called *Of Pandas and People* (P. Davis and D. H. Kenyon [Dallas: Haughton Publishing Co., 1989, 1993]). It is revealing and instructive to note that the 1987 draft of the *Pandas* book (then called *Biology and Origins*) used the word "creation" or "creationism" 272 times and the phrase "intelligent design" only 16 times, whereas the more recent (1993) edition of the book reversed that trend, with creation or creationism mentioned only 15 times but intelligent design used 103 times. In a similar vein, that same book uses the word "evolutionism" or "evolutionist" 97 times in the 1989 edition, and "Darwinist" or "Darwinism" only 9 times. The trend was reversed in their 1993 edition, with evolutionist/ism mentioned only 8 times and Darwinism occurring 133 times (data sent to me by Dr. Eugenie Scott, NCSE; see also the supplemental report on the origin of *Of Pandas and People* written by Dr. Barbara Forrest for use in the 2005 trial of *Tammy Kitzmiller et al. v. Dover Area School District* on the web at: http://creationismstrojanhorse.com/Forrest_Articles.html).

   The following quotations from both the earlier and later version of the book are revealing and help put to rest the argument that ID is different from creationism (emphasis is mine):

   > *Creation* means that various forms of life began abruptly through the agency of an *intelligent Creator* with their distinctive features already intact—fish with fins and scales, birds with feathers, beaks, and wings, etc. (*Biology and Origins,* 1987).
   >
   > *Intelligent design* means that various forms of life began abruptly through an *intelligent agency,* with their distinctive features already intact—fish with fins and scales, birds with feathers, beaks, and wings, etc. (*Pandas* 1993, 2nd edition, pp. 99–100).

   In a draft of a third edition of the *Pandas* book (eventually published in 2007 as *The Design of Life: Discovering Signs of Intelligence in Biological Systems,* with new authors William Dembski and Jonathan Wells), the manuscript of chapter 6 contained the following version, used as evidence in the Dover, Pennsylvania, trial (emphasis mine).

   > *Sudden emergence* holds that various forms of life began with their distinctive features already intact, fish with fins and scales, birds with feathers and wings, animals with fur and mammary glands.

   This version replaces ID with "sudden emergence" and significantly removes any reference to a specific designer or agent. This wording was removed before the book was published. See Eugenie Scott and Nicholas Matzke, "Biological Design in Science Classrooms," *Proceedings of the National Academy of Sciences* 104, suppl. 1 (2007): 8869–76. These word changes are not remotely related to science or science education; they are obvious shifts in tactics only.

3. See N. J. Matzke and P. R. Gross, "Analyzing Critical Analysis: The Fallback Antievolutionist Strategy," in E. C. Scott and G. Branch, eds., *Not in Our Classroom: Why Intelligent Design Is Wrong for Our Schools* (Boston: Beacon Press, 2006).
4. A document referred to as the "Wedge Document" warrants discussion here because it clearly outlines the real desire and agenda of the ID creationists. The document outlines a three-pronged attack on science education in the United States promoted by former law professor Phillip E. Johnson and the Discovery Institute's Center for Science and Culture. The strategy consists of three phases to be implemented over a period of years: Phase I: Scientific Research, Writing, and Publicity; Phase II: Publicity and Opinion-making; Phase III: Cultural Confrontation and Renewal. Clearly there is no concern with education or science here; this is an attempt to affect opinion and culture, ending with "cultural confrontation" and "renewal," and to insert one religious view into public schools.

   Taken verbatim from the Wedge Document is this chilling explanation of how the strategy's proponents view science and how they hope to end it: "If we view the predominant materialistic science as a giant tree, our strategy is intended to function as a wedge that, while relatively small, can split the trunk when applied at its weakest points." The second of their twenty-year goals is "to see intelligent design theory as the dominant perspective in science." Not satisfied even with this, their longer-term objective is "to see design theory permeate our religious, cultural, moral and political life." Thus, their concern for public science education simply does not exist; this is a ploy to set the stage for inserting their own brand of religion into all aspects of our lives. See Barbara Forrest and Paul R. Gross, *Creationism's Trojan Horse: The Wedge of Intelligent Design* (New York: Oxford University Press, 2007). For a condensed version, see Barbara Forrest's article, "The Wedge at Work: How Intelligent Design Creationism Is Wedging Its Way into the Cultural and Academic Mainstream," in *Intelligent Design and Its Critics,* edited by Robert T. Pennock (Cambridge, MA: MIT Press, 2001). The full text of the Wedge Document can be found on the NCSE web site at www.ncseweb.org/creationism/general/wedge-document.
5. Informal survey conducted by the National Science Teachers Association, 2005, see "Survey Indicates Science Teachers Feel Pressure to Teach Nonscientific Alternatives to Evolution," on the web at http://nsta.org/about/pressroom.aspx?id=50377.
6. See N. J. Matzke and P. R. Gross, "Analyzing Critical Analysis: The Fallback Antievolutionist Strategy," in E. C. Scott and G. Branch, eds., *Not in Our Classroom: Why Intelligent Design Is Wrong for Our Schools* (Boston: Beacon Press, 2006).

## CHAPTER 10. IS THERE EVIDENCE SUPPORTING INTELLIGENT DESIGN?

Epigraph. Paul Nelson is a fellow of the Discovery Institute's Center for Science and Culture, the main organization behind the ID movement. This quote is from an interview with Nelson and several others under the heading "The Measure of Design" in *Touchstone Magazine* 17, no. 6 (2004): 64–65.

1. This was the term used by Judge William Overton to point out the "either/or" fallacy promoted by creationists in the Arkansas "Balanced Treatment" case (see decision by Judge Jones in the Kitzmiller case, note 11).
2. See "The Flagellum Unspun: The Collapse of Irreducible Complexity," by Ken Miller, Brown University, at: http://millerandlevine.com/km/evol/design2/article.html.
3. For discussions that shed light on chance (providence) as an integral part of God's creation, see David Bartholomew, *God, Chance and Purpose: Can God Have It Both Ways?* (New York: Cambridge University Press, 2008); Fraser Watts, ed., *Creation: Law and Probability* (Minneapolis, MN: Fortress Press, 2008); and Elizabeth Johnson, "Does God Play Dice? Divine Providence and Chance," *Theological Studies* 57 (1996): 3–16 (also available online at http://www.aaas.org/spp/dser/03_Areas/evolution/perspectives/Johnson_1996.shtml).

   Elizabeth Johnson notes, "Since God and the world are in process together, not only does chance not threaten divine control over the universe, as it does in the classical model, but chance positively enriches divine experience (p. 11)." She goes on to say: "If this is the kind of universe created by the Holy Mystery who is God, then faith can affirm that God works not only through the deep regularities of the laws of nature but also through chance occurrence which has its own, genuinely random integrity. God uses chance, so to speak, to ensure variety, resilience, novelty, and freedom in the universe, right up to humanity itself (p. 15)."

   Similarly, Australian priest Denis Edwards (*The God of Evolution: A Trinitarian Theology* [Mahwah, NJ: Paulist Press, 1999]) writes "it would seem that a degree of randomness is essential for the creation of a universe anything like the one we have. God can be understood as the divine artist, achieving the divine purposes by working creatively and adventurously through the laws of nature and through chance (p. 124)."
4. Robert T. Pennock, *Tower of Babel: The Evidence against the New Creationism* (Cambridge, MA: MIT Press, 1999), p. 292.
5. One example is this quote from Galileo Galilei: "I do not feel obliged to believe that the same God who has endowed us with sense, reason, and intellect has intended [us] to forgo their use." (Letter to Madame Christina of Lorraine, Grand Duchess of Tuscany, 1615, in *Discoveries and Opinions of Galileo,* translated, with an introduction and notes, by Stillman Drake [New York: Random House, 1957], p. 183.)
6. Referring to evolutionary theory as "Darwinism" is intentional on the part of anti-evolutionists. It glosses over the fact that we have learned a huge amount about evolutionary biology in the 150 years since Darwin proposed the idea of natural selection (see "Let's Get Rid of Darwinism," by Olivia Judson, July 15, 2008, *New York Times,* online at www.judson.blogs.nytimes.com/2008/07/15/lets-get-rid-of-darwinism). It is also easier to attack a person rather than his or her ideas, especially meritorious ideas that have withstood the test of time. Additionally, the "-ism" is used to imply a dangerous ideology (compare to words such as Marxism and Nazism) and thereby confuse teachers and students (see Eugenie C. Scott and Glenn Branch, "Don't Call

It 'Darwinism,'" *Evolution: Education and Outreach* 2 [2009]: 90–94). The use of "Darwinism" is reminiscent of the time when flat-earth believers used the word "Newtonian" or "Newtonist" to refer to anyone who believed in a round earth. See also chapter 9, note 2, on the intentional decrease in the use of the words "evolutionism" or "evolutionist" and simultaneous increase in use of "Darwinist" or "Darwinism" in two editions of the ID book *Of Pandas and People*.

7. Theistic evolution has been described as the theological stance that "God creates diversity through evolution," a view compatible with a large number of religious organizations and religious leaders for whom scriptural creation stories are typically interpreted as being allegorical or metaphorical rather than literal. Historically, both Jews and Christians considered the idea of creation as being allegorical long before any of the writings of Charles Darwin. See also my introduction, note 7.

8. If evolution were really so evil, wouldn't you expect the people who know the most about it, namely evolutionary biologists, to be really bad people? But I have not found that to be the case.

9. As was pointed out by Judge John E. Jones III in his ruling on the 2005 Dover, Pennsylvania, trial, where the court concluded "that the religious nature of ID would be readily apparent to an objective observer, adult or child" (*Kitzmiller v. Dover* decision, p. 24 [see note 11]; see also chapter 9, note 2, and notes 6 and 10 in this chapter).

10. Perhaps my largest complaint about the ID movement is that it is dishonest. The major players are Christians, funded by evangelical Christian think tanks such as the Discovery Institute (DI) in Seattle (although the DI rejects being characterized as a Christian organization), representing predominantly Christian constituencies (especially pro-Christian school boards) in their attacks, and usually represented by Christian lawyers. Yet in the courtroom, and in their written attacks on science or other intellectual activities, they try hard to hide this fact. Judge Jones, in his 2005 decision, said about the Dover school board members "it is ironic that several of these individuals, who so staunchly and proudly touted their religious convictions in public, would time and again lie to cover their tracks and disguise the real purpose behind the ID Policy" (*Kitzmiller v. Dover* decision, p. 137 [see note 11]).

    Why would anyone do that? What would cause a person to be so disingenuous? The reason is simple: This nation will not allow any group to force its religious views on any other group, and that's what is happening here. So the IDC advocates masquerade as concerned citizens arguing for "fair and balanced treatment" in the schools and try to hide the fact that they are religiously motivated. I cannot imagine God ever being served well by dishonesty for any reason.

11. The entire decision of Judge Jones in the case of *Kitzmiller v. Dover Area School District* (400 F.Supp.2d 707 [M.D. Pa. 2005]), which includes a concise history of creationism and the ID movement, can be found on the web site of the NCSE and also at http://msnbcmedia.msn.com/i/msnbc/sections/news/051220_kitzmiller_342.pdf.

12. As one example, Brandeis University scholars Marc Z. Brettler and Bernadette J. Brooten, in a December 20, 2005, press release titled "Biblical Scholars Laud

Intelligent Design Ruling," voiced their strong support for Judge Jones's decision on theological grounds, calling it "a victory for the millions of Christians and Jews who do not use their faith to hinder the advancement of knowledge and who do not read the Bible literally." Similar support was voiced by the Vatican in January 2006 and again in 2009 (see appendix).

13. A teacher at Frazier Mountain High School, in Lebec, California, attempted to teach a class titled "Philosophy of Intelligent Design," thinking (incorrectly) that by calling it philosophy she could avoid the issue of separation of church and state. The case was quickly settled out of court.
14. See Glenn Branch and Eugenie C. Scott, "The Latest Face of Creationism," *Scientific American* 300, no. 1 (2009): 92–99.
15. Dr. Barbara Forrest's analysis of the bill and its creationist roots can be found at: http://lasciencecoalition.org//docs/Forrest_UpdatedAnalysis_SB_733_6.5.08.pdf.

## CHAPTER 11. HUMAN ARROGANCE

1. See J. D. Miller, E. C. Scott, and S. Okamoto, "Public Acceptance of Evolution," *Science* 313 (2006): 765–66.
2. In the translated words of the Papal Condemnation of 1633: "The proposition that the Sun is the center of the world and does not move from its place is absurd and false philosophically and formally heretical, because it is expressly contrary to Holy Scripture. The proposition that the Earth is not the center of the world and immovable but that it moves, and also with a diurnal motion, is equally absurd and false philosophically and theologically considered at least erroneous in faith." (See http://www.law.umkc.edu/faculty/projects/ftrials/galileo/condemnation.html for a summary of the trial compiled by Douglas O. Linder, who cites Georgio de Santillana, *The Crimes of Galileo* [Chicago: University of Chicago Press, 1955], pp. 306–10.)
3. As with so many "well-known" historical events, the Galileo incident and its aftermath are far more complex than what I have described here. See Galileo's own writings in *Discoveries and Opinions of Galileo,* translated, with an introduction, by Stillman Drake (New York: Anchor Books, 1957); Drake's translation of Galileo's letter to the grand duchess Christina is particularly illuminating with regard to science and religion. For a view of the Galileo incident that is more sympathetic to the Catholic Church than what I have presented, see Jerome J. Langford's *Galileo, Science, and the Church* (Ann Arbor: University of Michigan Press, 1992). For an in-depth analysis of the events following the trial over the next 359 years, see Maurice A. Finocchiaro's *Retrying Galileo,* 1633–1992 (Berkeley: University of California Press, 2007).
4. David Joravsky, *The Lysenko Affair* (Chicago: University of Chicago Press, 1986).
5. The documentary *Flock of Dodos,* by Randy Olson (2006), while mostly a critique of the ID movement, does a nice job of pointing out the inabilities of many scientists to communicate well without coming off as arrogant or condescending.

## CHAPTER 12. IN THE BEGINNING

1. For a list of the differences between Genesis 1 and 2, see Kenneth C. Davis, *Don't Know Much about the Bible* (New York: Eagle Book/Wm. Morrow and Co., 1998). For an overview of the theological implications of these differences, see chapter 1 in Eugene Peterson's *Christ Plays in Ten Thousand Places: A Conversation in Spiritual Theology* (Grand Rapids, MI: Wm. B. Eerdmans Co., 2005).
2. See Joan Roughgarden, *Evolution and Christian Faith: Reflections of an Evolutionary Biologist* (Washington, D.C.: Island Press, 2006).
3. This longing for a unity with all of creation is evident in Celtic Christian traditions of spirituality but largely lost or ignored elsewhere in the Christian world; see two books by J. Philip Newell—*Listening for the Heartbeat of God: A Celtic Spirituality* (Mahwah, N.J.: Paulist Press, 1997) and *The Book of Creation: An Introduction to Celtic Spirituality* (Mahwah, N.J.: Paulist Press, 1999).
4. Denis Edwards, *The God of Evolution: A Trinitarian Theology* (Mahwah, N.J.: Paulist Press, 1999), p. 124. According to the author, the writing of this book was stimulated in part by the 1996 conference on biological evolution and divinity sponsored by the Center for Theology and the Natural Sciences (Berkeley) and the Vatican Observatory in Castelgandalfo, Italy.

## CHAPTER 13. THE UNNECESSARY CHOICE

1. See Christine Garwood, *Flat Earth: The History of an Infamous Idea* (New York: MacMillan, 2007).
2. Gregory Koukl, "Does It Matter if We Evolved?" www.str.org/site/News2?page=NewsArticle&id=5210.
3. See www.theonion.com/content/node/39512.
4. Terry Eagleton, *Reason, Faith, and Revolution: Reflections on the God Debate* (New Haven, CT: Yale University Press, 2009), p. 37.

## EPILOGUE

Epigraph. This widely circulated quote probably did not originate with Einstein (see R. Keyes, *The Quote Verifier* [New York: St. Martin's Press, 2006], citing research by Alice Calaprice). Regardless of its origin, I like it.

1. My description of God permeating science in the same way that God permeates all of life does not answer all of our questions, of course. Where, for example, does intercessory prayer at the individual level fit in, and what sort of divine activity is assumed by such a practice? These questions are, regrettably, beyond the scope of this book and beyond my realm of knowledge.

APPENDIX

1. Sources for estimates of the number of Christians overall and within each denomination include: [T] *Time Almanac* (2002), which also forms the basis of the numbers used by the Ontario Consultants on Religious Tolerance web site (www.religious tolerance.org/); [Y] *Yearbook of American and Canadian Churches* (2007), published by the National Council of Churches USA, with some data repeated on their web site (www.ncccusa.org/members/index.html) (see also data compiled by the Hartford Institute for Religion Research, Hartford Seminary, at http://hirr.hartsem.edu/denom/homepages.html); [A] *Major Religions of the World Ranked by Number of Adherents* (online at www.adherents.com); [US] reports of the U.S. Census Bureau (under Population: Religion, Table 74, "Religious Composition of U.S. Population: 2007"—see http://census.gov/compendia/statab/tables/09s0074.pdf and http://census.gov/compendia/statab/tables/09s0075.pdf); [W] Wikipedia, based largely on Larry Diamond, Marc F. Plattner, and Philip J. Costopoulos, eds., *World Religions and Democracy* (Baltimore: Johns Hopkins University Press, 2005).

   See also Alister E. McGrath, *Christianity: An Introduction* (New York: Wiley-Blackwell, 2006); John R. Hinnells, *The Routledge Companion to the Study of Religion* (New York: Routledge, 2005); David Barrett et al., *World Christian Encyclopedia: A Comparative Survey of Churches and Religions in the Modern World* (New York: Oxford University Press, 2001); *Your Guide to the Religions of the World*, BBC World Service (www.bbc.co.uk/); David Gibbons, *Faiths and Religions of the World: The History, Culture, and Practice of Beliefs* (San Diego: Thunder Bay Press, 2007); and the CIA World Factbook, United States Government Central Intelligence Agency, at www.cia.gov/library/publications/the-world-factbook/geos/xx.html#People.
2. These percentages are from the Ontario Consultants on Religious Tolerance (OCRT) web site at www.religioustolerance.org/. Other estimates vary widely; for example, David Gibbons estimates 52% Protestant and only 24% Roman Catholic for the United States (see Gibbons 2007 [note 1, *above*], p. 22). The OCRT web site contains data from a variety of polls at www.religioustolerance.org/ev_publi.htm and a link to a list of what various denominations believe at: www.religioustolerance.org/ev_denom.htm.
3. The numbers I am using for the size (membership) of each group in the United States were obtained by averaging the high and low numbers from the sources given in note 1 (*above*). Numbers in parentheses are ranges of the estimates when different numbers were reported. Not all sources included all groups or denominations.
4. Additional statements from a wide variety of religious groups (Christian and non-Christian) that support the acceptance of evolutionary biology as being compatible with their faith can be found on the web site of the National Center for Science Education (www.ncseweb.org). These organizations include: American Jewish Committee, American Jewish Congress, American Scientific Affiliation, Central Conference of American Rabbis, Episcopal Church, Presbyterian Church USA, Lexington

Alliance of Religious Leaders, Lutheran World Federation, Roman Catholic Church, United Church Board for Homeland Ministries, Unitarian Universalist Association, United Methodist Church, and others. See also the NCSE document "Voices for Evolution."

5. Wikipedia (under "List of Christian denominations by number of members") gives 25 million as the membership of Calvary Chapel and 15 million for the Vineyard. These are overestimates. Calvary does not keep membership numbers and there are only approximately 1100 Calvary Chapels in the United States, the largest of which has perhaps 10–15,000 members (attendees), the smallest of which may be only 20–30 families (personal communication, Calvary Chapel Outreach Fellowship Office, April 2009). The 3.3 million is based on my arbitrarily using a figure of 3,000 attendees as an average for each of the 1100 churches.

# Glossary

**agnostic, agnosticism.** The philosophical point of view that certain statements or claims (usually those that deal with mystical, metaphysical, theological, or religious issues) are not knowable or are impossible to prove or disprove.

**artificial selection.** Human-influenced selection by breeding for selected desired traits or combinations of traits.

**atheism, atheist.** The rejection of theism (belief in God or gods); either a positive affirmation of the nonexistence of a deity or, more simply, the rejection of theism.

**creationist, creationism.** The belief that life, Earth, and the universe were created in their original form by a deity; often this is assumed to have occurred in the relatively recent past (*see* Young-earth creationism). Sometimes used to refer more specifically to the religiously motivated rejection of evolutionary biology.

**creation science.** The attempt by creationists to establish a scientific basis for belief in the Genesis account of creationism and/or attempt to disprove accepted scientific theories on the age of the earth, cosmology, and biological evolution.

**Darwinism.** An unclearly defined term usually used in a pejorative sense for concepts related to evolution by natural selection (as proposed by Darwin). For the most part, scientists do not use the term, as it conveys no information (similar to referring to someone who studies gravity as a Newtonist, for example). (See note 6 in chapter 10.)

**deist, deism.** The belief that God (or gods) created the known universe but no longer interacts with its operation. To a deist, the existence and nature of God are derived solely from reason; supernatural events (miracles, prophesies, divine revelation) are typically rejected.

**evangelical.** From the Greek *eu* (true, or good) and angel (messenger), and originally referring to "good news," an evangelist or evangelical was someone spreading the Christian faith, either by word or by example. Current usage varies; often it is meant to imply more conservative, revivalist, or fundamentalist leanings. (See note 4 in chapter 5.)

**evolution, evolutionist.** The process of biological change over time (descent with modification), encompassing small-scale evolution (changes in gene frequency in a population from one generation to the next) and large-scale evolution (the descent of different species from a common ancestor over many generations); also used to refer to the science of studying evolution. "Evolutionist" is sometimes used to refer to someone who understands, studies, or advocates for the teaching of evolution.

**evolutionary theory.** The overarching body of explanatory and causative theory that includes the facts, theories, and hypotheses about evolutionary biology.

**fact.** A verifiable observation or piece of information; an object or event that is demonstrably real. Also used commonly to mean "something for which we have so much evidence that it is no longer questioned."

**Flat-Earth creationism.** Form of creationism espousing belief in a flat Earth and basing that belief on certain passages in the Bible; a movement now assumed to be largely extinct in the United States.

**fundamentalist.** Often used pejoratively to imply someone lacking in serious theological education or with a narrow or inflexible viewpoint. In Christianity it usually refers to persons with a conservative theological position and to a movement that arose mostly in the United States and Great Britain at the start of the twentieth century, where in response to the encroachment on religion by "modernism," fundamentalists adopted and affirmed a set of basic (fundamental) beliefs.

**hypothesis.** In science, a hypothesis is a *testable* suggested explanation, based on previous observations, for a phenomenon or observable event.

**inerrant, inerrancy.** Without error. As concerns biblical interpretation, the position that the Bible is totally without error and free from all contradictions, including the historical and scientific parts. This is distinct from biblical *infallibility* (sometimes called *limited inerrancy*), which generally holds the Bible to be inerrant on issues of faith and practice but not on issues of history and science.

**intelligent design (ID).** The assertion that some features of the universe and of living systems are too complex to have arisen by the laws of nature and are therefore the result of planning by an intelligent agency (referred to as the designer); largely an outgrowth of "creation science" (which still exists) but with modifications.

**irreducible complexity.** An argument put forth by some advocates of intelligent design that some biological systems are too complex to have arisen (evolved) over time via the modification of simpler or less-complete predecessors.

**law.** In science, a statement that describes a particular behavior or set of behaviors in the natural world. Because the term is so often associated with physics, it is sometimes used synonymously with "physical law." Such "laws" refer to principles that are thought to be universal in nature.

**literalism, literalist.** In terms of biblical interpretation, literalism is the strict adherence to the written words of the Bible, not allowing for (or denying) the existence of allegory, parable, or metaphor in interpretation of scripture.

**macroevolution.** Large evolutionary changes, including speciation and extinction, usually (though not always) occurring over long periods of time and across separated gene pools.

**materialistic, materialism.** Concern with, interest in, or valuing of material things over spiritual or cultural values; someone is a materialist if they are an adherent of materialism.

**microevolution.** Small evolutionary changes, usually changes in gene frequency in a population or within a species.

**naturalistic, naturalism.** A naturalist is anyone who is a student of natural history—the scientific study of plants and animals. The word "naturalism" can refer to a wide variety of topics in philosophy and science. Of interest to our topic here, "naturalism" is sometimes used as a synonym for deism (*see above*), implying original creation of the world by God or gods but without further intervention; it is also used (rarely) to imply pure materialism (*see above*) or to refer to the philosophy that all phenomena labeled as supernatural or mystical are either false or are not different from natural phenomena.

**natural selection.** A nonrandom process of selection (differential survival) over successive generations, where inherited traits make it more likely for an organism to survive long enough to reproduce than for other organisms that lack the trait(s).

**progressive creationism.** A form of creationism that attempts to accommodate some aspects of modern science (e.g., mainstream geology and cosmology to allow for the known age of the earth and solar system) by postulating that new forms of life have continued to be created by divine intervention (as opposed to evolution) over time.

**rule.** In biology, a rule is an observed pattern that seems to be widespread in nature (e.g., Bergmann's rule, Cope's rule; see chapter 6).

**science.** From the Latin word *scientia* (knowledge), the systematic effort to discover new information—using controlled methods, observable physical evidence, experimentation, and the testing of falsifiable hypotheses—to increase human understanding of how the natural world works.

**specified complexity.** An argument put forth by ID proponents (primarily William Dembski) that certain complex features or systems in nature display a pattern of being both "specified" and also very complex; an argument for or indication of design.

**theism, theist.** Belief in the existence of one or more deities (gods). Thus, a monotheist believes in one god, a polytheist believes in multiple gods. Theism differs from deism in that theists hold that God remains actively interested in and continues to operate in the world.

**theistic evolution.** The philosophy that certain religious teachings, writings, and doctrines are not in conflict with evolutionary theory. Theistic evolution itself is not a theory; rather, it is the view that religious teachings about creation and scientific theories (specifically evolutionary theory) need not be viewed as contradictory.

**theory.** In science, a theory is a well-supported body of explanatory statements, including facts, laws, and hypotheses, that explains observations and can be used to make testable predictions. Scientific theories are frameworks into which the observed data fit.

**Young-Earth creationism.** A form of creationism that holds that the world and everything in it was created a very short time ago (usually 6,000 to 10,000 years) in the same form in which it is seen today.

# Recommended Further Reading

Ayala, Francisco J. 2006. *Darwin and Intelligent Design*. Minneapolis, MN: Fortress Press.

Baker, Catherine. 2007. *The Evolution Dialogues: Science, Christianity, and the Quest for Understanding,* 2nd edition. Washington, DC: American Association for the Advancement of Science.

Collins, Francis S. 2006. *The Language of God: A Scientist Presents Evidence for Belief*. New York: Free Press.

Comfort, Nathaniel C., ed. 2007. *The Panda's Black Box: Opening Up the Intelligent Design Controversy*. Baltimore: Johns Hopkins University Press.

Dowd, Michael. 2008. *Thank God for Evolution: How the Marriage of Science and Religion Will Transform Your World*. New York: Viking.

Falk, Darrel R. 2004. *Coming to Peace with Science: Bridging the Worlds between Faith and Biology*. Downer's Grove, IL: InterVarsity Press.

Forrest, Barbara, and Paul R. Gross. 2007. *Creationism's Trojan Horse: The Wedge of Intelligent Design*. New York: Oxford University Press.

Futuyma, Douglas J., ed. 1999. "Evolution, Science, and Society: Evolutionary Biology and the National Research Agenda. New Brunswick, NJ: Office of the University Publications, Rutgers, The State University of New Jersey. (View a PDF at www.rci.rutgers.edu/~ecolevol/fulldoc.pdf.)

Giberson, Karl W. 2008. *Saving Darwin: How to Be a Christian and Believe in Evolution*. New York: HarperOne.

Gomes, Peter J. 1996. *The Good Book: Reading the Bible with Mind and Heart*. New York: William Morrow (in particular chapter 15, "The Bible and Science").

Humes, Edward. 2007. *Monkey Girl: Evolution, Education, Religion, and the Battle for America's Soul*. New York: Ecco/HarperCollins.

Miller, Kenneth R. 1999. *Finding Darwin's God: A Scientist's Search for Common Ground between God and Evolution*. New York: Cliff Street Books/HarperCollins.

———. 2008. *Only a Theory: Evolution and the Battle for America's Soul*. New York: Viking (Penguin Group, USA).

National Academy of Sciences and Institute of Medicine. 2008. *Science, Evolution, and Creationism*. Washington, DC: The National Academies Press.

Peters, Ted, and Martinez Hewett. 2003. *Evolution from Creation to New Creation: Conflict, Conversation, and Convergence*. Nashville, TN: Abingdon Press.

———. 2006. *Can You Believe in God and Evolution? A Guide for the Perplexed*. Nashville, TN: Abingdon Press.

Pigliucci, Massimo. 2002. *Denying Evolution: Creation, Scientism, and the Nature of Science*. Sunderland, MA: Sinauer Associates, Inc.

Roughgarden, Joan. 2006. *Evolution and Christian Faith: Reflections of an Evolutionary Biologist*. Washington, DC: Island Press.

Ruse, Michael. 2000. *Can a Darwinian Be a Christian? The Relationship between Science and Religion*. New York: Cambridge University Press.

———. 2005. *The Evolution-Creation Struggle*. Cambridge, MA: Harvard University Press.

Sager, Carrie, ed. 2008. *Voices for Evolution*, 3rd ed. Berkeley, CA: National Center for Science Education, Inc. Online at http://ncseweb.org/voices.

Scott, Eugenie. 2008. *Evolution vs. Creationism: An Introduction*, 2nd ed. Westport, CT: Greenwood Press.

Scott, Eugenie, and Glenn Branch, eds. 2006. *Not in Our Classrooms. Why Intelligent Design Is Wrong for Our Schools*. Boston: Beacon Press.

Young, Matt, and Taner Edis, eds. 2006. *Why Intelligent Design Fails: A Scientific Critique of the New Creationism*, rev. ed. New Brunswick, NJ: Rutgers University Press.

Zimmer, Carl. 2009. *The Tangled Bank: An Introduction to Evolution.* Greenwood Village, CO: Roberts and Company.

# Helpful Web Sites

Understanding Evolution
www.evolution.berkeley.edu

The Clergy Letter Project
www.butler.edu/clergyproject/rel_evol_sun.htm

National Center for Science Education
www.ncseweb.org/

PBS: Evolution *(a web site to accompany the PBS television series)*
www.pbs.org/wgbh/evolution/

National Academy of Sciences *(section on Evolution)*
www.nationalacademies.org/evolution

National Science Teachers Association
www.nsta.org/, especially "Evolution Resources" at
www.nsta.org/publications/evolution.aspx

American Association for the Advancement of Science,
Dialogue on Science, Ethics, and Religion *(included is a free study guide to the AAAS booklet* The Evolution Dialogues*)*
www.aaas.org/spp/dser/

*The Stanford Encyclopedia of Philosophy, edited by Edward N. Zalta*
"Creationism," by Michael Ruse,
http://plato.stanford.edu/entries/creationism/

Talk Origins Archive
www.talkorigins.org/

The Center for Theology and the Natural Sciences
www.ctns.org/

# Index

African Methodist Episcopal Church (AME), 10, 118
agnosticism, defined, 151
Alabama, 77
Albert, Bruce, 137n1
Allen's rule, 43
Amish, 121–22
Anabaptists, 10, 121–22
ancestry, common, 48–49, 50
animals, breeding of, 49, 65
Anthony, Susan B., 81
Aquinas, Thomas, Saint, 113
arrogance, 5, 81–85
artificial selection, 49, 151
Assemblies of God, USA, 10, 122
astrology, 33
astronomy, 83–84, 96
astrophysics, 82
atheism, 4, 5, 58, 66–67, 73, 74; defined, 151
atoms/atomic theory, 40, 81
Augustine, Saint, 24, 113, 134n6

bacteria, flagellum of, 72
Baptist Churches, 115–16
Barry, A. L., 121
Behe, Michael, 70, 71, 72, 76, 138n5
Bellarmine, Roberto, 137n2
Bergmann's rule, 43
Berra, Yogi, 39
Bible: and creationism and intelligent design, 85; and earth as flat, 21, 22–23, 84; and evolution, 88–90, 121, 126; and science, 17, 18; study of, 99–100
Bible, books of: Acts, 74; 1 Chronicles, 134n4; 1 Corinthians, 89, 90; Ecclesiastes, 99; Ephesians, 89, 90; Ezekiel, 15; Isaiah, 55; John, 89; Jonah, 74; Joshua, 74; Leviticus, 74; Mark, 24, 27, 28, 29, 135n3; Matthew, 1, 5, 15, 135n4; Proverbs, 31, 47; Psalms, 9, 21, 81, 134n4; Revelation, 15, 22

biblical inerrancy, 29; and conflicts with science, 35; defined, 153; and earth as flat, 21; and evolutionary theory, 51, 56; as misunderstanding, 23, 24; and morality, 58
biblical literalism, 5, 29; and conflicts with science, 18–19, 35; defined, 153; and earth as flat, 21; and evolutionary theory, 51, 56; as misunderstanding, 23, 24; promotion of, 59
biology, 43, 47, 52, 100–101
Bonhoeffer, Dietrich, 59–60
Boyle's gas law, 42
Brethren in Christ, 121–22
Bush, George W., 77

Calvary Chapel, 10, 125–26, 127
Calvary Chapel Fort Lauderdale, 126–28
cell theory, 40, 72, 75
Center for Science and Culture, 65, 142n4
Chafin, Dewey, 135n2
Chambers, Oswald, 99
chance, 71, 73–74, 83, 104
change, 116; and creationism, 55; in evolutionary theory, 44, 48–49, 50, 65; and intelligent design, 70, 71; scientific view of, 41
Christianity, 99; acceptance of science in, 9–12, 58–59, 111–28; criticism of, 36; and earth as flat, 23, 25; and evangelicalism, 136n4; and evolution, 6, 51–52; and intelligent design, 76
Church of Christ, 10, 125
Church of God (Cleveland), 10, 123
Church of God in Christ, 10, 123, 126
Church of Jesus Christ of Latter-day Saints, 112
Church of the Brethren, 122
cloning, 77–78
Cobb County, Georgia, 57
Coleman, Randy, 126
Combined Anabaptist Groups, 121–22
complexity, 70–73; irreducible, 71–72, 153; specified, 73, 155
Congregationalism, 119–20
Copernicanism, 52, 83
Cope's rule, 43
Coyne, Jerry, 50
creationism, 2; arrogance of, 82, 85; characteristics of, 55–61; defined, 151; and Genesis, 60; and intelligent design, 64; and microevolution, 65; mistakes of, 59; pillars of, 58, 67; progressive, 70, 154; and trial of Galileo, 84; young-earth, 70
Creation Museum, 139n2

# Index

creation science, 60–61, 151
critical analysis, 67, 77, 78

Darwin, Charles, 44, 49, 50, 143n6
Darwinism, 94, 143n6, 151
Davis, P., 141n2
Dawkins, Richard, 35
Dembski, William, 73, 76, 138n5, 141n2
*Design of Life, The* (Dembski and Wells), 141n2
Discovery Institute, 65, 67, 77, 142n4
diversity/diversification, 48, 89, 96
Dobson, James, 77
Dover, Pennsylvania, case, 67, 72, 76–77
Doyle, Arthur Conan, 39

Eagleton, Terry, 95
earth, 93; age of, 55, 64, 66, 70; experienced as flat, 33; as flat in Bible, 21–25, 29, 31–32, 83, 84, 152; as flat in creationism, 56; and geocentrism, 55, 83; scientific view of, 24–25, 31–32, 51
Eastern Orthodox Church, 111
education, 74, 139n4; and creationism, 57, 59, 85; and creation science, 60–61; and empirical knowledge, 24–25; and evolutionary theory as religion, 58; and intelligent design, 66–67, 76–78, 85, 142n4
Edwards, Denis, 89
Einstein, Albert, 31, 33, 44, 103
electron, 81–82
Episcopal Church, 10, 114–15
Eratosthenes, 133n1
evangelicalism, 34, 94, 95, 115, 116, 126, 152; defined, 136n4, 152
Evangelical Lutheran Church in America, 10, 120–21
evolution, 40; and Christian worldview, 6; and complexity of life, 70–73; as conflicting with faith, 51–52, 82; debate over mechanisms of, 44, 70; defined, 152; evidence for, 50–51; and facts, 41; and natural selection, 44; non-selective mechanisms of, 50; and religious neutrality, 75; theistic, 144n7, 155; as threat, 82; through intermediate structures, 71–72
evolutionary theory, 47–52; as atheistic worldview, 75–76; Christian acceptance of, 9–12, 58–59, 111–28; and creationism, 56; as in crisis, 57; defined, 152; as flawed, 66, 67, 69; and God, 96; as religion, 58, 67; study of, 100–101

Facchini, Fiorenzo, 113
fact, 39, 40–42, 44–45, 152

faith, 58; as beyond realm of science, 35; as coexisting with science, 93–94; as complementing science, 100; as distinct from science, 6; evolution as conflicting with, 51–52, 82; and fundamentalism, 89–90; and physical evidence, 59; and scientific testing, 136n3
Fellowship Church, 126
First Amendment, 64, 67, 75
Flat Earth Society, 22, 23, 29, 31–32
flood, story of, 16, 56
Focus on the Family, 77
fossils, 56, 70
fundamentalism, 5, 89–90, 152

Galileo, 9, 34, 83–84, 93, 111, 143n5
Garwood, Christine, 22, 25
Genesis, 15–16, 17, 18, 51; as basis of creationism, 55; composition of, 88, 140n12; creation in, 87–90; and creationism, 60; creation of humankind in, 82; and evolution, 88–89; and microevolution, 65; truth of, 56, 58, 59, 60
genetics, 43, 45, 65, 84
germ theory, 40, 43
God, 6, 17, 18; and chance, 74; creation by, 82, 87–89; as designer, 71, 72, 73, 76; and diversity through evolution, 144n7; as Elohim, 88; as employing evolution, 75; and evolution, 51, 96, 103–4; exclusion of, 70, 71, 75; existence of, 84; of gaps, 59–60, 73; image of, 82; as indecisive or fickle, 73; and intelligent design, 63, 64, 65; and knowledge, 59; limits on, 18, 74; and oneness of creation, 89; and rainbows, 15–16; and science, 84, 95; as trickster, 56; as Yahweh, 88
Goethe, Johann Wolfgang von, 15, 47, 69
Gomes, Peter, 5, 55
gradualism, 50
Graham, Billy, 11, 51, 138n7
Grand Canyon, 94, 96
Gravity/gravitational theory, 40, 41, 43, 44, 52, 95
Greek Orthodox Church, 10, 113–14
Gregorios, Paulos Mar (metropolitan), 113–14

half-a-wing argument, 71–72
Ham, Ken, 126, 139n2
Hardy-Weinberg law, 43
Haught, John, 11
Hawaiian Islands, 73
Hawking, Stephen, 1, 63
heliocentrism, 40, 52, 58, 96. *See also* sun/sunlight
Hensley, George, 28

Hitchens, Christopher, 35
Hutterites, 122
hypothesis: defined, 42, 153; testing of, 32, 33, 39, 43; and theory, 44

intelligent design, 24, 60, 63–67; arrogance of, 82, 85; and creationism, 64; defined, 153; as dishonest, 144n10; evidence in support of, 69–78; and Galileo, 84; positions in, 70; and science, 76
International Church of the Foursquare Gospel, 10
International Circle of Faith, 10, 123–24
Isay, David, 135n2

Jehovah's Witnesses, 112
Jesus Christ, 24, 27, 89, 100
John Paul (pope), 136n5
John Paul II (pope), 11, 51, 112
Johnson, Charles K., 133n2
Johnson, Phillip E., 66, 76, 142n4
Jones, John E., III, 76–77, 144nn9, 10

Kansas, 57, 139n4
Kenyon, D. H., 141n2
kinds, variation within, 65
*Kitzmiller v. Dover,* 144n9
knowledge, 59, 70

Lakewood Church (Houston), 126
Lamarck, Jean-Baptiste, 44
law, 39, 42–43, 153
Lebec, California, 77
Lennon, John, 87
Life Church (Edmond, Okla.), 126
Louisiana Family Forum, 77
Louisiana Science Education Act, 77
Lutheran Church–Missouri Synod, 10, 121
Lutheranism, 120–21
Lysenko, Trofim, 84

macroevolution, 65, 154
Manning, Brennan, 27
McCartney, Paul, 87
megachurches, 12, 126–27
Mendelian genetics, 43, 84

Mennonites, 121–22
Methodism, 116–18
Michigan, 77
microevolution, 65, 154
Miller, Ken, 59
Mississippi, 77
Missouri, 77
Mohler, R. Albert, Jr., 115–16
morality, 58, 94
Murphy, Nancey, 52, 63

National Baptist Convention, USA, 10, 116
National Center for Science Education, 77
National Science Teachers Association, 66
natural selection, 70, 116; characteristics of, 49; and complexity of life, 71; and Darwin, 44, 50; defined, 154; study of, 45
Nelson, Paul, 69, 70
New Apostolic Church, 10
Newton, Isaac, 9, 16–17, 43, 44, 96
Non-Denominational Evangelicalism, 125–28
North Point Community Church (Alpharetta, Georgia), 126

*Of Pandas and People* (Davis and Kenyon), 141n2
Oklahoma, 77
Open Brethren, 121–22
Overton, William, 143n1

Paul (the apostle), 89
Peacocke, Arthur, 11
Pennock, Robert, 74
Pentecostalism, 122–24, 126
physics, 18, 42, 43
Pius XII (pope), 51
Polkinghorne, John, 11
Potter's House (Dallas), 126
Presbyterian Church USA, 2, 10, 118–19
Priest, Ivy Baker, 21
prism: and appreciation for nature, 19; and Bible, 18; and biblical literalism, 23; and morality in Bible, 52; Newton's proof concerning, 16–17; and scientific discovery, 51, 96
Protestantism, 111

# Index

Quakers, 121–22

rainbow, 15–19, 23, 75, 96. *See also* prism
randomness, 71, 104
Ravasi, Gianfranco, 113
Religious Society of Friends, 121–22
Restorationism, 124–25
Roman Catholic Church, 10, 34, 83–84, 111, 112–13
Rowbotham, Samuel Birley, 133n3, 134n5
rule, 39, 43, 154

Saddleback Church (Lake Forest, California), 126–27
science, 2; and atheism, 75; and Bible, 17, 18; Christian acceptance of, 9–12, 111–28; and creationism, 59; defined, 154; and earth as flat, 24–25, 32; and faith, 6, 35, 93–94, 100; and God, 84, 95, 96, 101; knowledge refined in, 70; method of, 4–5, 32, 36; misunderstanding of, 34; and naturalistic phenomena, 34, 35, 36; and oneness of creation, 90; perceived as anti-religious, 5; progress through, 32–33; questions not addressed by, 34; repeatability in, 33; as self-correcting, 33; study of, 100–101
scientists, beliefs of, 4, 9, 35, 83, 94
Second Baptist (Houston), 127
selection. *See* artificial selection; natural selection
separation, of church and state, 64
Seventh-Day Adventists, 10, 124–25
Shenton, Samuel, 134n5
Signs Following movement, 27–29
Sims, Bennett J., 93, 114–15
Skehan, Jim, 61
Smith, Chuck, 126
snake dancers/handlers, 28–29, 31, 83
South Carolina, 77
Southern Baptist Convention, 10, 115–16, 127
Soviet Union, 84
speciation, 50, 65
species, 50, 65, 73, 83
Stand to Reason (web site), 94
stars, 56
Stenger, Victor J., 35
sun/sunlight, 16, 17, 58, 96. *See also* heliocentrism; prism
Superstring Theory, 90
Supreme Court, U.S., 64

teach the controversy, 24, 66
testability, 42, 71, 135n1
theory, 39–45; current, 39; definition of, 43–45, 155; and doubt, 40; and fact, 40. *See also* cell theory; germ theory; gravity/gravitational theory; heliocentrism
Theory of Everything, 90
Thomas Aquinas, Saint, 113
Thwaites, William, 58

Unitarian Universalists, 112
United Church of Christ, 10, 119–20
United Methodist Church, 10, 116–18
United States, 2, 11, 57, 58, 59
United States Supreme Court, 64
Utah, 77

variation, 49, 65, 71
Vineyard, The, 10, 126
Voliva, Wilbur Glenn, 134n5

Wade, Bernie L., 123–24
Wedge Document, 142n4
Wedge Strategy, The, 66
Wells, Jonathan, 141n2
West Angeles Cathedral (Los Angeles), 126
Willow Creek Community Church (Illinois), 126